Language, images and artificial intelligence

语言、图像与人工智能

主　编　徐雪涛
副主编　赵　雄　张　艳
参　编　张艳丽　龙佳红
　　　　饶玉芳　黄　浩

武汉理工大学出版社
·武汉·

图书在版编目(CIP)数据

语言、图像与人工智能:英文/徐雪涛主编. —武汉:武汉理工大学出版社,2016.12
ISBN 978-7-5629-5388-3

Ⅰ.①语… Ⅱ.①徐… Ⅲ.①人工智能-研究-英文 Ⅳ.①TP18

中国版本图书馆 CIP 数据核字(2016)第 270707 号

项目负责:陈军东　彭佳佳　　　　　　责任编辑:徐　扬
责任校对:彭佳佳　　　　　　　　　　封面设计:芳华时代
出版发行:武汉理工大学出版社
地　　址:武汉市洪山区珞狮路 122 号
邮　　编:430070
网　　址:http://www.wutp.com.cn 理工图书网
E-mail:chenjd@whut.edu.cn
经 销 者:各地新华书店
印 刷 者:荆州市鸿盛印务有限公司
开　　本:787×960　1/16
印　　张:11.25
字　　数:250 千字
版　　次:2016 年 12 月第 1 版
印　　次:2016 年 12 月第 1 次印刷
定　　价:28.00 元

凡购本书,如有缺页、倒页、脱页等印装质量问题,请向出版社发行部调换。
本社购书热线电话:(027)87515798　87165708

Preface

It is only in comparatively recent years that linguistics has begun to be studied and taught. Because it is a comparative newly-emerged science, there is a tendency to regard it as a difficult and esoteric subject. Linguistics, defined as the scientific study of language, is intricately to analyse and to study, because language is associated with human Intelligence; Human Intelligence is based on language and language on concepts. Only when the way how concepts come into being and function in human brain have been revealed, can we nearly be able to announce that the mystery of human Intelligence is revealed.

Artificial Intelligence (AI) is to help machines find solutions to complex problems in a more human-like fashion. This generally involves imitating human intelligence, and applying them as algorithms in a computer friendly way. Although AI links with a lot of fields such as: Psychology, Cognition, Biology and Philosophy, etc., but we believe, it mainly links with linguistics, more specifically, the concepts of language. We are not having no good computer programmers, but having no good linguistic theories.

Computer is fundamentally well suited to performing mechanical computations, using fixed programmed rules. Computers can perform simple monotonous tasks efficiently, which humans are ill-suited to. For more complex problems, things get more difficult… Unlike humans, computers have trouble understanding specific situations, and adapting to new situations. Artificial Intelligence aims to improve machine behaviour in tackling such complex tasks.

linguistics study is allowing us to understand our intelligence. Humans have an unequaled capability to problem-solving, based on abstract thought and conceptual reasoning. Artificial Intelligence can recreate this process. But to date, all the traits of human intelligence have not been captured and applied to spawn intelligent artificial creatures.

The potential applications of Artificial Intelligence are abundant. They stretch from the military for auto-control and target identification, to the entertainment industry from computer games to robotic pets.

Devoting to research on linguistics and to understand the nature of language intelligence can help to remould computers that exhibit true intelligence. A truly intelligent computer would be more flexible and would engage in the kind of "thinking" that people really do. An example is vision. A array of sensors combined with systems for interpreting the data may produce the kind of pattern recognition that we take for granted as seeing and understanding what we see. In fact, writing software that can recognize subtle differences in objects (such as those we perceive in the faces of two people) is very difficult. Actually, differences between faces of two people that we can perceive without deliberate effort, we believe, owes much to our integrated proportional judgement among mutli-cognitive domains of faces comparation rather than to massive amounts of data of faces and careful guidelines for a system of artificial intelligence to recognize. Computer tries to imitate true intelligence, you can't copy intelligence it if you don't know how language works.

Several editors have made contributions to the publication of this book. The selection from publications is a group of effort. Though we aimed to present a comprehensive volume of introductory reading on the relationships between linguistics and artificial intelligence, the lack of sources for choice to the area of study covered by this book. So this book is open for suggestions and criticism we hope to hear from the readers of this book.

Xu Xuetao
October, 11, 2015

Table of Contents

Unit 1　Language Defined ……………………………………… (1)

Unit 2　Language and Thought …………………………………… (8)

Unit 3　Properties of Language …………………………………… (21)

Unit 4　The Origin of Language …………………………………… (36)

Unit 5　What Is Linguistics ……………………………………… (58)

Unit 6　The Object of Linguistics ………………………………… (67)

Unit 7　Nature of the Linguistic Sign …………………………… (73)

Unit 8　Some Basic Concepts in Linguistics …………………… (80)

Unit 9　Phonology ………………………………………………… (95)

Unit 10　Phonetic Distinctive Features ………………………… (101)

Unit 11　Form and Meaning in Natural Languages …………… (112)

Unit 12　Semantics and Semantic Theory ……………………… (132)

Unit 13　Pragmatics ……………………………………………… (150)

Unit 14　Psycholinguistics ……………………………………… (159)

Unit 15　Artificial intelligence and computer modeling ……… (165)

参考文献 …………………………………………………………… (172)

Unit 1

Language Defined

> This is an excerpt from Language: *An Introduction to the Study of speech* by Edward Sapir(1884—1939), an American linguist. With his student Benjamin Lee Whorf (1897—1941) developed the Sapir-Whorf hypothesis, arguing that the limits of Language restrict the scope of possible thought and that every language recognizes peculiar sets of distinctions-e. g. Eskimo and its rich vocabulary for different kinds of snow. In this excerpt, he, having distinguished speech from other functions of man (such as walking) and from mere imitation of things, tries to give a serviceable definition of language. He also discusses the nature of speech, and the relation between language and thought.

Speech is so familiar a feature of daily life that we rarely pause to define it. It seems as natural to man as walking, and only less so than breathing. Yet it needs but a moment's reflection to convince us that this naturalness of speech is but an illusory feeling. walking is an inherent, biological function of man. Not so language. The process of acquiring speech is, in sober fact, a different sort of thing from the process of learning to walk.

Walking is an organic, an instinctive, function (not, of course, itself an instinct); speech is a non-instinctive, acquired, "cultural" function. It is a purely historic heritage of the group, the product of long-continued social usage. It varies as all creative effort varies.

Therefore, language is a human and non-instinctive method of communica-

ting ideas, emotions, and desires by means of a system of voluntarily produced symbols. These symbols are, in the first instance, auditory and produced by organs of speech. There is no discernible instinctive basis in human speech as such, however much instinctive expressions and the natural environment may serve as a stimulus for the development of certain elements of speech, however much instinctive tendencies, motor and other, may give a predetermined range or mold to linguistic expression. Such human or animal communication, if "communication" it may be called, as is brought about by involuntary, instinctive cries is not, in our sense, language at all.

I have just referred to the "organs of speech," and it would seem at first blush that this is tantamount to an admission that speech itself is an instinctive, biologically predetermined activity. We must not be misled by the mere term. There are, properly speaking, no organs of speech; there are only organs that are incidentally useful in the production of speech sounds. The lungs, the larynx, the palate, the nose, the tongue, the teeth, and the lips, are all so utilized, but they are no more to be thought of as primary organs of speech than are the fingers to be considered as essentially organs of piano-playing. Speech is not a simple activity that is carried on by one or more organs biologically adapted to the purpose. It is an extremely complex and ever-shifting network of adjustments—in the brain, in the nervous system, in the articulating and auditory organs—tending towards the desired end of communication. The lungs developed, roughly speaking, in connection with the necessary biological function known as breathing; the nose, as an organ of smell, the teeth, as organs useful in breaking up food. If, then, these and other organs are being constantly utilized in speech, it is only because any organ, once existent and in so far as it is subject to voluntary control, can be utilized by man for secondary purposes. Physiologically, speech is a group of overlaid functions. It gets what service it can out of organs and functions, nervous and muscular, that have come into being and are maintained for very different ends than its own.

It is true that physiological psychologists speak of the localization of speech in the brain. This can only mean that the sounds of speech are localized in the auditory tract of the brain, or in some circumscribed portion of it, precisely as other classes of sounds are localized, and that the motor processes involved in speech are localized in the motor tract precisely as are all other impulses to special motor activities. In the same way control is lodged in the visual tract of the brain over all those processes of visual recognition involved in reading. Naturally the particular points or clusters of points of localization in the several tracts that refer to any element of language are connected in the brain by paths of association, so that the outward, or psycho—physical, aspect of language is of a vast network of associated localizations in the brain and lower nervous tracts, the auditory localizations being without doubt the most fundamental of all for speech. However, a speech-sound localized in the brain, even when associated with the particular movements of the "speech organs" that are required to produce it, is very far from being an element of language. It must be further associated with some element or group of elements of experience, say a visual image or a class of visual images or a feeling of relation, before it has even rudimentary linguistic significance. This "element" of experience is the content or "meaning" of the linguistic unit; the associated auditory, motor, and other cerebral processes that lie immediately back of the act of speaking and the act of bearing speech are merely a complicated symbol of or signal for these "meanings", of which more anon. We see therefore at once that language as such is not and can't be definitely localized, for it consists of a peculiar symbolic relation—physiologically an arbitrary one—between all possible elements of consciousness on the one hand and certain selected elements localized in the auditory, motor, and other cerebral and nervous tracts on the other. If language can be said to be definitely "localized" in the brain, it is only in that general and rather useless sense in which all aspects of consciousness, all human interest and activity, may be said to be "in the brain." Hence, we have no recourse but to accept language as a fully formed functional system within man's psychic or "spiritual" constitution. We cannot define it as an entity in psycho—physical terms alone, however much the psycho—physical basis is essential to its functioning in the individual.

From the psychologist's point of view we may seem to be making an unwarrantable abstraction in desiring to handle the subject of speech without constant and explicit reference to that basis. However, such an abstraction is justifiable. We can profitably discuss the intention, the form, and the history of speech, precisely as we discuss the nature of any other phase of human culture—say art or—religion as an institutional or cultural entity, leaving the organic and psychological mechanisms back of it as something to be taken for granted. Accordingly, it must be clearly understood that this introduction to the study of speech is not concerned with those aspects of physiology and of physiological psychology that underlie speech. Our study of language is not to be one of the genesis and operation of a concrete mechanism; it is, rather, to be an inquiry into the function and form of the arbitrary systems of symbolism that we term languages.

I have already pointed out that the essence of language consists in the assigning of conventional, voluntarily articulated, sounds, or of their equivalents, to the diverse elements of experience. The word "house", is not a linguistic fact if by it 1S meant merely the acoustic effect Produced on the ear by its constituent consonants and vowels, pronounced in a certain order; nor the motor processes and tactile feelings which make up the articulation of the word; nor the visual perception on the Part of the hearer of this articulation; nor the visual perception of the word "house" on the written or printed page; nor the motor processes and tactile feelings which enter into the writing of the word; nor the memory of any or all of these experiences. It is only when these, and possibly still other, associated experiences are automatically associated with the image of a house that they begin to take on the nature of a symbol, a word, an element of language. But the mere fact of such an association is not enough. One might have heard a particular word spoken in an individual house under such impressive circumstances that neither the word nor the image of the house ever recur in consciousness without the other becoming present at the same time. This type of association does not constitute speech. The association must be a purely symbolic one; in other words, the word must denote, tag off, the image, must have no other significance than to serve as a counter to refer to it whenever it is necessary or convenient to do so. Such an association, voluntary and, in a sense, arbitrary as it is, demands a considerable exercise of self-conscious attention. At least to begin with, for habit soon makes the association nearly as automatic as any and more rapid than most.

But we have traveled a little too fast. Were the symbol "house",—whether an auditory, motor, or visual experience or image—attached but to the single image of a particular house once seen, it might perhaps by an indulgent criticism, be termed an element of speech, yet it is obvious at the outset that speech so consti-

tuted would have little or no value for purposes of communication. The world of our experiences must be enormously simplified and generalized before it is possible to make a symbolic inventory of all our experiences of things and relations and this inventory is imperative before we can convey ideas. The elements of language, the
Chinese House
symbols that ticket off experience, must therefore be associated with whole groups, delimited classes, of experience rather than with the single experiences themselves. Only so is communication possible. To be communicated it needs to be referred to a class which is tacitly accepted by the community as an identity. Thus, the single impression which I have had of a particular house must be identified with all my other impressions of it. The particular experience that we started with has now been widened so as to embrace all possible impressions or images that sentient beings have formed or may form of the house in question. This first simplification of experience is at the bottom of a large number of elements of speech, the so-called proper nouns or names of objects. It is, essentially, the type of simplification which underlies, or forms the crude subject of, history and art. But we can't be content with this measure of reduction of the infinity of experience. We must cut to the bone of things, we must more or less arbitrarily throw whole masses of experience together as similar enough to warrant their being looked upon—mistakenly, but conveniently—as identical. This house and that house and thousands of other phenomena of like character are thought of as having enough in common, in spite of obvious differences of detail, to be classed under the same heading. In other words, the speech element "house" is the symbol, first and foremost, not of a single perception, nor even of the notion of a particular object, but of a "concept"—of a convenient capsule of thought that embraces thousands of distinct experiences and that is ready to take in thousands more. If the significant elements of speech are the symbols of concepts, the actual flow of speech may be interpreted as a record of the setting of these concepts into mutual relations.

Words and Expressions

sober	['səubə]	adj.	头脑清醒的,冷静的,严肃的;朴素的,素净的
heritage	['heritidʒ]	n.	遗产;继承物;传统;文化遗产;传承
circumscribed	['sɜːkəmskraibd]	adj.	[医]局限的;受限于有限空间的
		v.	在…周围画线;划定…范围;限制;限定
variability	[ˌvɛəriəbiləti]	n.	变化性,易变,变化的倾向;变率
involuntary	[in'vɔləntəri]	adj.	非故意的;非自愿的,不随意的;不由自主的;无意识的;偶然的
assignable	[ə'sainəbəl]	adj.	可分配的,可指定的;不可忽视,可转让的
discernible	[di'sɜːnəbəl]	adj.	可识别的;可辨别的
predetermined	[ˌpriːdɪ'tɜːmɪnd]	v.	预先确定的(predetermine 的过去式和过去分词)
		adj.	形容词预定的;预先确定的;预先决定的
at first blush			猛一看,乍看
tantamount	['tæntəmaunt]	adj.	相等的,相当的,等值的;等价的;
larynx	['lærɪŋks]	n.	喉;喉头;咽喉;喉部
palate	['pælət]	n.	〈解〉腭;味觉,嗜好;审美眼光,鉴赏力
articulating	[ɑː'tikjulitɪŋ]	n.	表达,表述 v. 清楚地表
physiological	[ˌfɪzɪə'lɔdʒɪkl]	adj.	生理学的;生理的
psychologist	[saɪ'kɔlədʒɪsts]	n.	心理学研究者,心理学家
lodge	[lɔdʒ]	v.	存放,暂住,埋入,(权利、权威等)归属
rudimentary	[ˌruːdɪ'mentri]	adj.	基本的,初步的,发育全的,未成熟的,退化的
anon	[ə'nɒn]	adv.	不久以后
unwarrantable	[ʌn'wɒrəntəbl]	adj.	无正当理由的;无法辩解的;无法律依据的;不合法的,不允许的;不能

		保证的；不能承认的；毫无道理
explicit [ik'splisit]	adj.	明确的,清楚的；直言的；详述的；不隐瞒的
justifiable ['dʒʌstifɛəbl]	adj.	有理由的；正当的；入情入理；说得过去
institutional [,insti'tju:ʃnl]	adj.	由来已久的；习以为常的；公共机构的；
entity ['entəti]	n.	实体；实际存在物；本质
indulgent [in'dʌldʒənt]	adj.	放纵的,纵容的；宽容的；任性的
perception [pə'sepʃn]	n.	知觉；觉察(力)，观念,(农作物)收获,感知

Questions for Discussion and Review

1. What is the basic difference between speech and other forms of functions of man like walking?

2. How do you understand "house" as a symbol?

3. Does concepts build on the basis of similarities or on the basis of differences?

4. Similarity and Difference, which one is absolutely, which one is relatively? Why?

5. Similarity, Difference and Diversity, what are their differences?

Unit 2

Language and Thought

> The question has often been raised whether thought is possible without speech; further, if speech and thought be not but two facets of the same psychic process. The question is all the more difficult because it has been hedged about by misunderstanding.

In the first place, it is well to observe that whether or not thought necessitates symbolism, that is speech, the flow of language itself is not always indicative of thought. We have seen that the typical linguistic element labels a concept. It does not follow from this that the use to which language is put is always or even mainly conceptual. In ordinary life, we are not so much concerned with concepts as such as with concrete particularities and specific relations. When I say, for instance, "I had a good breakfast this morning," it is clear that I am not in the throes of laborious thought, that what I have to transmit is hardly more than a pleasurable memory symbolically rendered in the grooves of habitual expression. Each element in the sentence defines a separate concept or conceptual relation or both combined, but the sentence as a whole has no conceptual significance whatever. It is somewhat as though a dynamo capable of generating enough power to run an elevator were operated almost exclusively to feed an electric doorbell. The parallel is more suggestive than at first sight appears. Language may be looked upon as an instrument capable of running a gamut of psychic uses. Its flow not only parallels that of the inner content of consciousness, but parallels it on different levels, ranging from the state of mind that is dominated by particular images to that in which abstract concepts and their relations are alone at the focus of attention and which is ordinarily termed reasoning. Thus the outward form only of language is constant; its inner meaning, its psychic value or intensity, varies freely with attention or the selective interest of the mind, also, needless to say, with the mind's general development. From the point of view of language, thought may be defined as the highest latent or potential content of speech, the content that is obtained by interpreting each of the elements in the flow of language as possessed of its very fullest conceptual value. From this it follows at once that language and thought are not strictly coterminous. At best language can but be the outward facet of thought on the highest, most generalized, level of symbolic expression. To put our viewpoint somewhat differently, language is primarily a prerational function. It humbly works up to the thought that is latent in, that may eventually be read into, its classifications and its forms; it is not, as is generally but naively assumed, the final label put upon the finished thought.

Most people, asked if they can think without speech, would probably answer, "Yes, but it is not easy for me to do so. Still I know it can be done." Language is but a garment!

But what if language is not so much a garment as a prepared road or groove? It is, indeed, in the highest degree likely that language is an instrument originally put to uses lower than the conceptual plane and that thought arises as a refined interpretation of its content. The product grows, in other words, with the instrument, and thought may be no more conceivable, in its genesis and daily practice, without speech than is mathematical reasoning practicable without the lever of an appropriate mathematical symbolism. No one believes that even the most difficult mathematical proposition is inherently dependent on an arbitrary set of symbols, but it is impossible to suppose that the human mind is capable of arriving at or holding such a proposition without the symbolism. The writer, for one, is strongly of the opinion that the feeling entertained by so many that they can think, without language is an illusion. The illusion seems to be due to a number of factors. The simplest of these is the failure to distinguish between imagery and thought. As a matter of fact, no sooner do we try to put an image into conscious relation with another than we find ourselves slipping into a silent flow of words. Thought may be a natural domain apart from the artificial one of speech, but speech would seem to be the only road we know of that leads to it.

A still more fruitful source of the illusive feeling that language may be dispensed with in thought is the common failure to realize that language is not identical with its auditory symbolism. The auditory symbolism may be replaced, point for point, by a motor or by a visual symbolism (many people can read, for instance, in a purely visual sense, that is, without the intermediating link of an inner flow of the auditory images that correspond to the printed or written words) or by still other, more subtle and elusive, types of transfer that are not so easy to define. Hence the contention that one thinks without language merely because he is not aware of a coexisting auditory imagery is very far indeed from being a valid one. One may go so far as to suspect that the symbolic expression of thought may in some cases run along outside the fringe of the conscious mind, so that the feeling of a free, non-linguistic stream of thought is for minds of a certain type a relatively, but only a relatively, justified one. Psycho-physically, this would mean that the auditory or equivalent visual or motor centers in the brain, together with the appropriate paths of association, that are the cerebral equivalent of speech, are touched off so lightly during the process of thought as not to rise into consciousness at all. This would be a limiting case-thought riding lightly on the submerged crests of speech, instead of jogging along with it, hand in hand. The modern psychology has shown us how powerfully symbolism is at work in the unconscious mind. It is therefore easier to understand at the present time than it would have been twenty years ago that the most rarefied thought may be but the conscious counterpart of an unconscious linguistic symbolism.

One word more as to the relation between language and thought. The point of view that we have developed does not by any means preclude the possibility of the growth of speech being in a high degree dependent on the development of thought. We may assume that language arose pre-rationally-just how and on what precise level of mental activity we do not know-but we must not imagine that a highly developed system of speech symbols worked itself out before the genesis of distinct concepts and of thinking, the handling of concepts. We must rather imagine that thought processes set in, as a kind of psychic overflow, almost at the beginning of linguistic expression; further, that the concept, once defined, necessarily reacted on the life of its linguistic symbol, further linguistic growth. We see this complex process of the interaction of language and thought actually taking place under our eyes. The instrument makes possible the product, the product refines the instrument. The birth of a new concept is invariably foreshadowed by a more or less strained or extended use of old linguistic material; the concept does not attain to individual and independent life until it has found a distinctive linguistic embodiment. In most cases the new symbol is but a thing wrought from linguistic material already in existence in ways mapped out by crushingly despotic precedents. As soon as the word is at hand, we instinctively feel, with something of a sigh of relief, that the concept is ours for the handling. Not until we own the symbol do we feel that we hold a key to the immediate knowledge or understanding of the concept. Would we be so ready to die for "liberty," to struggle for "ideals," if the words themselves were not ringing within us? And the word, as we know, is not only a key; it may also be a fetter.

Language is primarily an auditory system of symbols. The motor aspect of speech is clearly secondary to the auditory. In normal individuals the impulse to speech first takes effect in the sphere of auditory imagery and is then transmitted to the motor nerves that control the organs of speech. The motor processes and the accompanying motor feelings are not, however, the end. They are merely a means and a control leading to auditory perception in both speaker and hearer. Communication is successfully effected only when the hearer's auditory perceptions are translated into the appropriate and intended thought. Hence the cycle of speech begins and ends in the realm of sounds. The concordance between the initial auditory imagery and the final auditory perceptions is the warrant of the successful issue of the process. Therefore, the typical of this process may undergo endless modifications or transfers into equivalent systems without thereby losing its essential formal characteristics.

The most important of these modifications is the abbreviation of the speech process involved in thinking. This has doubtless many forms, according to the structural or functional peculiarities of the individual mind.

The least modified form is that known as "talking to one's self." Here the speaker and the hearer are identified in a single person. More significant is the still further abbreviated form in which the sounds of speech are not articulated at all. To this belong all the varieties of silent speech, silent reading and of normal thinking. The auditory centers alone may be excited; or the impulse to linguistic expression may be communicated as well to the motor nerves that communicate with the organs of speech but be inhibited either in the muscles or at some point in the motor nerves themselves; or, possibly, the auditory centers may be only slightly affected, the speech process manifesting itself directly in the motor sphere. There must be still other types of abbreviation. How common is the excitation of the motor nerves in silent speech, in which no audible or visible articulations result, is shown by the frequent experience of fatigue in the speech organs, particularly in the larynx, after unusually stimulating reading or intensive thinking.

All the modifications so far considered are directly patterned on the typical process of normal speech. Of very great interest and importance is the possibility of transferring the whole system of speech symbolism into other terms than those that are involved in the typical process. This process, as we have seen, is a matter of sounds and of movements intended to produce these sounds. The sense of vision is not brought into play. But let us suppose that one not only hears the articulated sounds but sees the articulations themselves as they are being executed by the speaker. Clearly, if one can only gain a sufficiently high degree of adroitness in perceiving these movements of the speech organs, the way is opened for a new type of speech symbolism-that in which the sound is replaced by the visual image of the articulations that correspond to the sound. This sort of system has no great value for most of us because we are already possessed of the auditory-motor system of which it is at best but an imperfect translation, not all the articulations being visible to the eye. However, it is well known what excellent use deaf-mutes can make of "reading from the lips" as a subsidiary method of apprehending speech. The most important of all visual speech symbolism is, of course, that of the written or printed word, to which, on the motor side, corresponds the system of delicately adjusted movements which result in the writing or typewriting or other graphic method of recording speech. The significant feature for our recognition in these new types of symbolism, apart from the fact that they are no longer a by-product of normal speech itself, is that each element (letter or written word) in the system corresponds to a specific element (sound or sound-group or spoken word) in the primary system. Written language is thus a point-to-point equivalence, to borrow a mathematical phrase, to its spoken counterpart. The written forms are secondary symbols of the spoken ones—symbols of symbols—yet so close is the correspondence that they may, not only in theory but in the actual practice of certain eye-readers and, possibly, in certain types of thinking, be entirely substituted for the spoken ones. Yet the auditory-motor associations are probable always latent at the least, that is, they are unconsciously brought into play. Even those who read and think without the slightest use of sound imagery are, at last analysis, dependent on it. They are merely handling the circulating medium, the money, of visual symbols as a convenient substitute for the economic goods and services of the fundamental auditory symbols.

The possibilities of linguistic transfer are practically unlimited. A familiar example is the Morse telegraph code, in which the letters of written speech are represented by a conventionally fixed sequence of longer or shorter ticks. Here the transfer takes place from the written word rather than directly from the sounds of spoken speech. The letter of the telegraph code is thus a symbol of a symbol of a symbol. It does not, of course, in the least follow that the skilled operator, in order to arrive at an understanding of a telegraphic message, needs to transpose the individual sequence of ticks into a visual image of the word before he experiences its normal auditory image. The precise method of reading off speech from the telegraphic communication undoubtedly varies widely with the individual. It is even conceivable, if not exactly likely, that certain operators may have learned to think directly, so far as the purely conscious part of the process of thought is concerned, in terms of the tick-auditory symbolism or, if they happen to have a strong natural bent toward motor symbolism, in terms of the correlated tactile-motor symbolism developed in the sending of telegraphic messages.

Still another interesting group of transfers are the different gesture languages, developed for the use of deaf-mutes, of Trappist monks vowed to perpetual silence, or of communicating parties that are within seeing distance of each other but are out of earshot. Some of these systems are one-to-one equivalences of the normal system of speech; others, like military gesture-symbolism or the gesture language of the Plains Indians of North America (understood by tribes of mutually unintelligible forms of speech) are imperfect transfers, limiting themselves to the rendering of such grosser speech elements as are an imperative minimum under difficult circumstances. In these latter systems, as in such still more imperfect symbolism as those used at sea or in the woods, it may be contended that language no longer properly plays a part but that the ideas are directly conveyed by an utterly unrelated symbolic process or by a quasi-instinctive imitativeness. Such an interpretation would be erroneous. The intelligibility of these vaguer symbolisms can hardly be due to anything but their automatic and silent translation into the terms of a fuller flow of speech.

We shall no doubt conclude that all voluntary communication of ideas, aside from normal speech, is either a transfer, direct or indirect, from the typical symbolism of language as spoken and heard or, at the least, involves the intermediary of truly linguistic symbolism. This is a fact of the highest importance. Auditory imagery and the correlated motor imagery leading to articulation are, by whatever devious ways we follow the process, the historic fountain-head of all speech and of all thinking. One other point is of still greater importance. The ease with which speech symbolism can be transferred from one sense to another, from technique to technique, itself indicates that the mere sounds of speech are not the essential fact of language, which lies rather in the classification, in the formal patterning, and in the relating of concepts. Once more, language, as a structure, is on its inner face the mold of thought. It is this abstracted language, rather more than the physical facts of speech, that is to concern us in our inquiry.

There is no more striking general fact about language than its universality. One may argue as to whether a particular tribe engages in activities that are worthy of the name of religion or of art, but we know of no people that is not possessed of a fully developed language. The lowliest South African Bushman speaks in the forms of a rich symbolic system that is in essence perfectly comparable to the speech of the cultivated Frenchman. It goes without saying that the more abstract concepts are not nearly so plentifully represented in the language of the savage, nor is there the rich terminology and the finer definition of nuances that reflect the higher culture. Yet the sort of linguistic development that parallels the historic growth of culture and which, in its later stages, we associate with literature is, at best, but a superficial thing. The fundamental groundwork of language—the development of a clear-cut phonetic system, the specific association of speech elements with concepts, and the delicate provision for the formal expression of all manner of relations—all this meets us rigidly perfected and systematized in every language known to us. Many primitive languages have a formal richness, a latent luxuriance of expression, that eclipses anything known to the languages of modern civilization. Even in the mere matter of the inventory of speech the layman must be prepared for strange surprises. Popular statements as to the extreme poverty of expression to which primitive languages are doomed are simply myths. Scarcely less impressive than the universality of speech is its almost incredible diversity. Those of us that have studied French or German, or, better yet, Latin or Greek, know in what varied forms a thought may run. The formal divergences between the English plan and the Latin plan, however, are comparatively slight in the perspective of what we know of more exotic linguistic patterns. The universality and the diversity of speech lead to a significant inference. We are forced to believe that language is an immensely ancient heritage of the human race, whether or not all forms of speech are the historical outgrowth of a single pristine form. It is doubtful if any other cultural asset of man, be it the art of drilling for fire or of chipping stone, may lay claim to a greater age. I am inclined to believe that it antedated even the lowliest developments of material culture, that these developments, in fact, were not strictly possible until language, the tool of significant expression, had itself taken shape.

Words and Expressions

facets	['fæsits]	n.	(facet 的复数)方面,(事物,宝石)小平面,面
psychic	['saikik]	adj.	精神的;超自然的;灵魂的,心灵的
hedged about			束缚
necessitates	[ni'sesiteits]	v.	使…成为必要,需要(necessitate 的第三人称单数)
particularities	[pə,tikju'læritiz]	n.	特性(particularity 的名词复数)
throes	[θrəuz]	n.	剧痛(如分娩时的阵痛);挣扎
laborious	[lə'bɔːriəs]	adj.	费力的;勤劳的;辛苦的,艰苦的
rendered	['rendəd]	v.	使(render 的过去式和过去分词);放弃;表达;提供
as though			好像,仿佛;浑似
dynamo	['dainəməu]	n.	精力充沛的人;[物]发电机
exclusively	[ik'skluːsivli]	adv.	唯一地,专门地,特定地,专有地;排外地,仅仅地
suggestive	[sə'dʒestiv]	adj.	提示的,暗示的,提醒的,引起联

			想的,高度怀疑
gamut	['gæmət]	n.	全范围,全部;音阶;音域
psychic	['saikik]	adj.	精神的;超自然的;灵魂的,心灵的
outward form			外观形态
latent	['leitnt]	adj.	潜在的;潜伏的;休眠的;潜意识的
		n.	指纹,指印
coterminous	[kəʊ'tə:minəs]	adj.	有共同边界,毗连,(事物或看法)几乎一致
at best			至多,充其量;顶多
prerational	[pri'ræʃənel]	adj.	前理性的,在智力发展以前的
plane	[plein]	n.	水平;平面;飞机;木工刨
		adj.	平的,平坦的
refined	[ri'faind]	adj.	精炼的;精制的;经过改良的;举止优雅的
conceivable	[kən'si:vəbl]	adj.	可想到的,可相信的,可想像的,可能的
genesis	['dʒenəsis]	n.	创始,起源,发生,成因
proposition	[ˌprɔpə'ziʃn]	n.	命题;建议;主张
		v.	提议;建议
inherently	[in'hiərəntli]	adv.	天性地,固有地,内在地;本质上
entertained	[ˌentə'teind]	v.	款待,招待(entertain 的过去式和过去分词);使欢乐
dispensed with			放弃省去
elusive	[i'lu:siv]	adj.	难以捉摸的,不易记住的,逃避的,难以找到,难懂的
submerged	[səb'mə:dʒd]	adj.	在水中的,淹没的,水下运动的
		v.	(使)潜入水中,淹没(submerge 的过去式和过去分词);完全掩盖,遮掩
crests	[krests]	v.	到达山顶(或浪峰);crest 的第三人称单数;到达洪峰,达到顶点
preclude	[pri'klu:d]	vt.	阻止,排除,妨碍,预防,防止,杜绝
rationally	['ræʃnəli]	adv.	讲道理地,理性地,合理地

invariably	[inˈvɛəriəbli]	adv.	总是；不变的
foreshadowed	[fɔːˈʃædəud]	v.	预示，是…的先兆（foreshadow 的过去式,过去分词）
invariably foreshadowed			总是埋下伏笔
strained	[streind]	adj.	紧张的,不友善的,态度勉强的
embodiment	[Imˈbɒdimənt]	n.	体现；化身；具体化
wrought	[rɔːt]	adj.	制造加工的,经装饰的;(金属)锤打成形的
		vt.	使发生了,造成了(尤指变化)
		v.	(使)工作(work 的过去式和过去分词);(使)运作；运转;(使)产生效果
map out			在地图上标出,筹划某事；详细提出某事；映射出
crushingly	[ˈkrʌʃiŋli]	adv.	(强调不好的特征)极其，决定性地
despotic	[diˈspɒtik]	adj.	专制暴君的,似暴君的,专横的,暴虐,豪强
precedents	[ˈpresidənt]	n.	前例,先例,判例,成例
		adj.	在前的,在先的,
fetter	[ˈfetə(r)]	n.	脚镣；束缚
		vt.	给…上脚镣,束缚
realm	[relm]	n.	领域,范围,王国,区域部门,(动植物)圈,带
concordance	[kənˈkɔːdəns]	n.	和谐性,一致性
modifications	[ˌmɒdəfiˈkeiʃnz]	n.	缓和(modification 的名词复数),限制,更改,改变,条款修订；契约修订；修改的内容；修饰作用
peculiarities	[piˌkjuːliˈæritiːz]	n.	古怪(peculiarity 的名词复数);怪癖,特色,奇形怪状;奇特之处,怪异之处,特色风貌,特别之处
inhibited	[inˈhibitid]	adj.	拘谨的；压抑的；不自在的
		v.	抑制；使拘束；阻止；使尴尬
excitation	[ˌeksaiˈteiʃən]	n.	<物>(原子获得高能量的)激发(过程),激动,(由刺激引起器

			官、组织等的)应激反应
adroitness	[ə'drɔit]	n.	熟练
subsidiary	[səb'sidiəri]	adj.	附带的,附属的,次要的,帮助的;补足的
		n.	附属事物,附属机构,子公司;附属者,附属品
tactile	['tæktail]	adj.	触觉的;触觉感知的;能触知的;有形的
trappist	['træpist]	n.	(天主教中)特拉普派(此派强调缄口苦修)
vow	[vau]	n.	誓言;郑重宣布
		vt. & vi.	起誓,发誓;郑重宣告
perpetual	[pə'petʃuəl]	adj.	永久的;不断的;无期限的;四季开花的
earshot	['iəʃɔt]	n.	听力所及的范围,可听距离
Plains Indians		n.	平原印第安人
unintelligible	[ˌʌnin'telidʒəbl]	adj.	莫名其妙的,模糊的,难以理解的
rendering	['rendəriŋ]	n.	翻译;表演;
grosser	['grəusə]	adj.	总的(gross 的比较级),粗俗的,易见的,粗鲁的
imperative	[im'perətiv]	adj.	必要的,不可避免的,命令的,[语]祈使的
		n.	必要的事,命令,规则;[语]祈使
quasi	['kweisai]	adj.	类似的,准的
erroneous	[irəuniəs]	adj.	错误的,不正确的,秕谬
intelligibility	[inˌtelidʒə'biləti]	n.	可理解性,可理解的事物
vaguer	[veigə]	adj.	模糊的(vague 的比较级),不清楚的,含糊的
intermediary	[ˌintə'mi:diəri]	n.	媒介,中间人,调解人,中间阶段
imagery	['imidʒəri]	n.	比喻,形象化的描述,意象
articulation	[aːˌtikju'leiʃn]	n.	清晰度,咬合,关节,发音
fountain-head		n.	根源;源地;源头
ease	[i:z]	n.	轻松,舒适;容易;安逸;不拘束,自在

mold [məʊld]	n.	模子;模式;类型;霉	
	vt.	塑造;浇铸;用模子做	
	vi.	对…产生影响,形成;发霉	
universality [ˌjuːnɪvɜːˈsæləti]	n.	普遍性;通用性;普适性;一般性	
lowliest [ˈləʊliːst]	adj.	低级的,最贫贱的,最低贱;卑微的	
savage [ˈsævɪdʒ]	adj.	未开化的;野蛮的;凶猛的;残忍的	
	n.	未开化的人,野蛮人;野兽	
terminology [ˌtɜːmɪˈnɒlədʒi]	n.	专门名词;术语,术语学;用辞	
nuance [ˈnjuːɑːns]	n.	细微差别;细微的表情,精微玄妙之处	
superficial [ˌsuːpəˈfɪʃl]	adj.	表面上的,肤浅的,缺乏深度的,一知半解的	
clear-cut	adj.	轮廓鲜明的;清晰的	
provision [prəˈvɪʒn]	n.	规定,条款;预备,准备,设备;供应品	
primitive [ˈprɪmətɪv]	adj.	原始的,落后的;[生物学]原生的	
	n.	原始人;早期的艺术家(作品),单纯的人	
latent [ˈleɪtnt]	adj.	潜在的;潜伏的;休眠的;潜意识的	
	n.	指纹,指印	
luxuriance [lʌgˈʒʊəriəns]	n.	茂盛,华丽;精美	
eclipse [ɪˈklɪps]	n.	[天](日、月)食;消失,黯然失色;漆黑	
inventory [ˈɪnvəntri]	n.	存货清单;财产目录,存货总值	
layman [ˈleɪmən]	n.	门外汉,外行;俗人;一般信徒,凡人	
myth [ˈmɪθ]	n.	神话,杜撰出来的人[事物],荒诞说法	
scarcely less		简直一样,简直相等	
divergence [daɪˈvɜːdʒəns]	n.	分叉,分歧,背离;离题	
inference [ˈɪnfərəns]	n.	推理;推断;推论	
heritage [ˈherɪtɪdʒ]	n.	遗产;继承物;传统;文化遗产	
pristine [ˈprɪstiːn]	adj.	太古的,原始的,原始状态的,纯	

		(质)朴的,纯洁的
asset ['æset]	n.	资产,财产,有用的东西,优点
lay claim to		声称…的权力, 据理力争
inclined to		偏向于,倾向于
antedate ['ænti‚deit]	v.	先于,早于,(在信、支票等上)到签日期

Questions for Discussion and Review

1. In the author's opinion, what is the relation between language and thought? What do you think of this opinion?

2. What do you think of the definition of language given by the author?

3. Non-sound animal communication, like the dances of bees, can transfer the originator of the flower message to others. Is it language, why?

Unit 3

Properties of Language

> The ability to use language is the most distinctive characteristic of human beings. In this article, Bolton, American Professor of English, analyzes the defining properties of human language and explains the intricate mechanisms involved in speech production and reception.

Language is so built into the way people live that it has become an axiom of being human. It is the attribute that most clearly distinguishes our species from all others; it is what makes possible much of what we do, and perhaps even what we think. Without language we could not specify our wishes, our needs, the practical instructions that make possible cooperative endeavor ("You hold it while I hit it"). Without language we would have to grunt and gesture and touch rather than tell. And through writing systems or word of mouth we are in touch with distant places we will never visit, people we will never meet, a past and a future of which we can have no direct experience. Without language we would live in isolation from our ancestors and our descendants, condemned to learn only from our own experiences and to take our knowledge to the grave.

Of course other species communicate too, sometimes in ways that seem almost human. A pet dog or cat can make its needs and wishes known quite effectively, notonly to others of its own species but to its human owner. But is this language? Porpoises make extremely complex sequences of sounds that may suggest equally complex messages,

But make wishs without language

but so far no way has been found to verify the suggestion. Chimpanzees have been taught several humanly understandable languages, notably AMESLAN (American Sign Language) and a computer language, but there has been heated debate whether their uses of these languages are like ours or merely learned performances of rather greater subtlety than those of trained circus animals. If the accomplishments of dolphins and chimpanzees remain open questions, however, there is no question but that human uses of language, both everyday and in the building of human cultures, are of a scope and power unequaled on our planet.

It seems likely that language arose in humans about a hundred thousand years ago. How this happened is at least as unknowable as how the universe began, and for the same reason: There was nobody there capable of writing us a report of the great event. Language, like the universe, has its creation myths; Modern linguists, like modern cosmologist, have adopted an evolutionary hypothesis. Somehow, over the millennia, both the human brain and those parts of the human body now loosely classed as the organs of speech have evolved so that speech is now a part of human nature. Babies start to talk at a certain stage of their development, whether or not their parents consciously try to teach them; Only prolonged isolation from the sounds of speech can keep them from learning.

Writing is another matter. When the topic of language comes up, our first thoughts are likely to be of written words. But the majority of the world's languages have never been reduced to writing, and illiteracy is a natural state; We learn to write only laboriously and with much instruction. This is hardly surprising, since compared with speech writing is a very recent invention-within the past 5,000 years. Still more recently there have been invented complex languages of gesture for use by and with people unable to hear or speak; these too must be painstakingly learned. What do the spoken, written, and sign languages have in common that distinguishes them from other ways to communicate?

1.Properties of Language

Perhaps the most distinctive property of language is that its users can create sentences never before known, and yet perfectly understandable to their hearers and readers. We don't have to be able to say "I've heard that one before!" in order to be able to say, "I see what you mean." And so language can meet our expressive needs virtually without limit, no matter how little we have read or heard before, or what our new experiences call on us to express. Another way of describing this property is to say that language is productive. We take this productivity for granted in our uses of language, but in fact it is one of the things that make human communication unique.

Less obvious is the fact that language is arbitrary: the word for something seldom has any necessary connection with the thing itself. English say *one*, *two*, *three*—but the Chinese say *yi*, *er*, *san*. Neither language has the "right" word for the numerals, because it is arbitrary. (It might seem that a dog's barking were equally arbitrary, as it might be translated into various languages as "Go away!" or "zou kai! "—but within the species the sound is universally understandable. A Chinese chow and a German shepherd understand each other without translation unlike speakers of Chinese and German.)

Even the sounds of a language are arbitrary. English can be spoken using only 36 significantly different sounds, and these are not all the same as the sounds needed to speak other language. These 36 sounds are in turn arbitrarily represented by 26 letters, some standing for two or more sounds, others overlapping. (Consider c, s, and k.) And the patterns into which these sounds, and indeed words, may be arranged are also arbitrary. In English we usually put an adjective before its noun-fat man; in French it's the other way around, *homme gros*. This patterning is the key to the productivity of language. If we use intelligible words in proper patterns, we can be sure of being understood by others who speak our language. Indeed, we seem to understand nonsense, provided it is fitted into proper patterns—the silly nonsense of pun, the impressive nonsense of much bureaucratese.

This ability to attach meaning to arbitrary clusters of sounds or words is like the use of symbolism in literature and art. The words one does not somehow represent the numeral, somehow embody its essence the way a three-sided plane figure represents the essence of triangularity. Rather, *one* merely stands for the prime numeral 1, giving a physical form to the concept, Thus the sound /wʌn/ spelled one, has a dual quality as a sound and as a concept. This can be seen from the fact that /wʌn/ spelled won, matches the identical sound to a wholly different concept. This feature of duality is both characteristic of and apparently unique in human communication, and so linguists use it as a test to distinguish language from other kinds of communication in which a sound can have only a single meaning. (Such sounds are called signs, to distinguish them from the symbols that are human words.)

Sounds can be made into meaningful combinations, such as language, only if they are first perceived as meaningfully distinct, or discrete. We can find an analogy in music. Musical pitch rises continuously without steps from the lowest frequency we can hear to the highest, sliding upward like the sounds of a siren. But most of music is not continuous; it consists of notes that move upwards in discrete steps, as in a scale. This is why we can talk about notes being the same or different, as we could not easily do if all possible tones from low to high were distributed along a continuous line. Similarly, in speech we can slide through all the vowels from "ee" in the front of the mouth to "aw" in the throat-but then how could we tell *key* from *kay* from *coo* from *caw*? Likewise we distinguish between *v* and *f*, so that view is different from few. But these distinctions are arbitrary. They are not even common to all language. For example, in German the letters *v* and *f* both represent the sound /f/, the letter *w* represents the sound /v/ and there is no sound /w/. What all languages do have in common, however, is the property of discreteness.

These four properties, or "design features," of language were set down to see how human language differs from animal communication. There are of course other design features, but these four (productivity, arbitrariness, duality, and discreteness) appear to be the most important.

Among the others:

Our language acts are specialized. speech is not necessary for breathing, nor is it the same as other sounds we make, such as a laugh or a cry of pain or fear. Of course, such sounds can communicate, but only by accident to those within earshot. Their main purpose is a reflexive one: they happen more or less involuntarily, like the jerk of a tapped knee.

Italian children grow up speaking Italian, Chinese children learn Chinese. *Human language is transmitted by the cultures we live in*, not by our parentage: if the Chinese infant is adopted or living in Italy, he or she will grow up speaking Italian. But a kitten growing up among human beings speaks neither Italian nor Chinese; it says meow. Its communication is determined by its genetic make up, not by its cultural context.

2. Non-languages

Other kinds of human communication are sometimes called body language, or kinesics. The way we use our bodies in sitting, standing, walking, is said to be expressive of things we do not say. It probably is, but that does not make it language. Body language lacks duality, in that it is not symbolic but rather a direct representation of a feeling; discreteness, in that there is no "alphabet" of distinctive movements or postures; and productivity, in that "original" expressions are likely not to be understood. Moreover, it appears to be only partly arbitrary, for the movement or posture is often selected by its "meaning" as representational, not arbitrary; "barrier signs" such as crossing one's arms or legs need no dictionary.

3. The Physiology of Speech

The discomfort of breathing during speech does not take place during actual speech. Our neural and biochemical makeup is in fact specially adapted so that we can sustain the speech act. Other animal species are equally adapted to their systems of communication, but none of them can be taught ours because ours is species-specific, a set of abilities that have evolved in humankind over a very long time. The evolution has included the most intricate adaptations of the body and its workings, particularly the neural system (above all, the brain); the motor system (especially the muscles that the neural system controls); and the sensory system (especially hearing, but also touch).

The speech act involves an input of meaning and an output of sound on the part of the speaker, the reverse on the part of the listener. But a great deal takes place between the input and the output, and it takes place in the brain. That means that the organ for thinking, the brain, is by definition the seat of language. And the brain is also the control center for the intricate virtuoso muscular performance we call speech, commanding the vocal activities and—most important—ensuring their coordination and sequencing.

The brain is not just an undifferentiated mass in which the whole organ does all of its tasks. The different tasks that the brain dose are localized, and in a more general way, the whole brain is lateralized. In most people, the right half (hemisphere) controls the left half of the body and vice-versa, and many brain functions are also lateralized. Language is one of them; it is localized in several areas of the left Hemisphere. The language centers are not motor control centers for the production of speech. Instead, they are "boardroom" in which decisions are made, decisions that motor control centers in both hemispheres of the brain implement by issuing the orders to the body. The orders are carried by electric impulses from the central nervous system (brain and spinal cord) into the peripheral nervous system (activating the muscles).

Wernicke's area lies in the left hemisphere of the brain, just above the ear. It takes its name from the German Carl Wernicke (1848—1905), who in 1874 showed that damage to that part of the brain leads to a disrupted flow of meaning in speech. A decade earlier the Frenchman Paul Broca (1824—1880) had shown that damage to another area of the left hemisphere, several inches further downward, led instead to disrupted pronunciation and grammar. There are also differences in the areas when it comes to receptive ability: damage to Broca's area does not much affect comprehension, but damage to Wernicke's area disrupts it seriously.

These differences suggest that the two chief language areas of the brain have functions that are distinct but complementary. It seems that the utterance gets its basic structure in Wernicke's area, which sends it on to Broca's area through a bundle of nerve fibers called the *arcuate fasciculus*. In Broca's area the basic structure is translated into the orders for the speech act itself, which go on to the appropriate motor control area for implementation. In reverse order, a signal from the hearing or the visual system (speech or writing) is relayed to Wernicke's area for decoding from language to linguistic meaning. Broca's area, which seems to write the program for the speech act, is not so important to listening or reading as Wernicke's area is.

All of this, naturally, is inferential: the evidence as we know it points to these conclusions, but no one has ever actually seen these brain activities taking place. The conclusions are also incredible. It is difficult to imagine all that activity for a simple "Hi!" But those conclusions are the simplest ones that will account adequately for the evidence.

All sound, whether a cat's meow, a runner's "Hi!", is a disturbance of the air in which it is produced. When the sound is speech it can be studied in terms of its production (articulatory phonetics), its physical properties in the air (acoustic phonetics), or its reception by the ear and other organs of hearing (auditory phonetics). The first of these is the easiest to study without special instruments, and it is the only one of the three that directly involves the motor system.

4. The Sensory System

In a way, the main sensory system of language, hearing, is the reverse of speech. Speech turns meaning into sound, while hearing turns sound into meaning. Speech encodes meaning as language in the brain, and the brain sends neural messages to the motor system for action; the motor system produces speech. Hearing turns the speech sounds back into neural messages which go to the brain where they are decoded into language and interpreted for meaning.

Sound, as we have seen, is a disturbance of the air—a kind of applied energy. The ear is designed to pick up and process that energy, often incredibly small amounts. The ear is good not only at amplifying small sounds but at damping loud ones, within limits: a very sudden, very loud noise, or even sound that is not sudden (if it is loud enough), can cause damage to the sensitive sound-gathering mechanisms of the ear, damage which if severe or prolonged can be permanent.

What we usually mean by *ear* is the appendage earrings hang from, but that is only the ear's most visible part. In fact it has three divisions: the outer ear, which extends into the eardrum; the middle ear; and the inner ear. The outer ear collects the sound, it through the ear canal, and focuses it on the ear drum. The ear drum is a tightly stretched membrane which is set into motion by the vibrations of sound energy; it is really a "drum" in reverse, for while the bass drum in a marching band converts the energy of motion (a blow from a drumstick) into sound waves, the ear drum converts sound waves into motion energy which is picked up in the middle ear. That motion is carried through the middle ear by three tiny bones; here weak sounds are amplified and very strong sounds are damped. The last of the three bones delivers the sound motion to a membrane called the oval window, which is smaller than the ear drum; the difference in size helps to concentrate the sound energy.

The oval window divides the middle ear from the inner ear. The inner ear is composed of several cavities in the bones of the skull, in one of these, the cochlea, the energy that arrived at the outer ear as sound, and is now motion, will be converted by a set of intricate organs into electrical impulses and fed into the central nervous system for delivery to the auditory center of the brain. The remaining steps in the process are then neural, not sensory.

The process here described, and our idea of hearing in general, relates to sound that reaches us from outside by conduction through the air, water, or other medium. But there is another way in which we can receive sound. A vibrating fork held against the skull will be "heard" by conduction through the bone itself, even if the ear hole is effectively plugged. Bone-conduction helps us monitor our own speech by providing continuous feedback; thus we can pick out our own words even when surrounded by loud conversation or noise. Bone-conduction has a different sound quality from air-conduction, which is why your voice sounds to you one way when you are speaking and another when you hear it played back from a tape. And bone-conduction can sometimes substitute for air-conduction, for example when a hearing aid "plays" sound waves directly into the bones of the skull.

5. Language and Cultural Relativism

Language is species-specific to humankind. By "humankind" we mean the genus Homo, species sapiens of this genus survives. Any smaller subdivisions, such as sex or race, may differ among themselves in other very visible ways, but the neural, motor, and sensory equipment necessary to language is common to all. Not that the equipment is identical; otherwise everyone would speak at about the same pitch. But racial, sexual, or individual differences in the shape and size of the nose and lips, or of the internal speech organs, do not override the structural similarity of the vocal organs among all human groups, and they definitely do not result in any functional differences. The members of any group, that is, have the vocal organs to articulate any human language with complete mastery. The same is true of other genetic factors: the intellectual ability to use language is the same in all the varieties of humankind and in all normal individuals.

That is not the same as saying that adult individuals can learn a foreign language as easily as they learned their own in childhood. The physiological habits of the speech organs are complex, and they are learned early. We observe that a native speaker of Chinese has difficulty with the sound of *r* in *very*, a native speaker of Japanese with the sound of /l/ in *hello*. That is because their native languages have given them no opportunity to practice those sounds. On the contrary, the languages have reinforced other sounds that tend to crop up when the Chinese speaker attempts English /r/ or the Japanese speaker English /l/. The problem, however, is one of habit and not heredity. An American of Chinese ancestry has no trouble with the sounds of English, including /r/, while a person of European ancestry raised to speak Chinese would.

Our virtuosity in our own language carries with it other commitments, some easily understandable and some less so. Speakers of English easily handle a system of pronouns that distinguishes among masculine (he), feminine (she), and neuter (it) forms. They may have trouble with a language like German, however, where the nouns, adjectives, and articles (equivalents of *the* and *a*) make a similar three-way distinction, often in apparent disregard of the sex of the noun—a maiden (*das Mädchen*) is neuter, and stays that way when she becomes a wife (*das Frau*)—or with a language like French which makes only a two-way distinction between masculine and feminine, so that *table* is feminine (*la table*) but *floor* is masculine (*le plancher*).

We should not rush to conclude, however, that the Germans and the French see sexual characteristics in inanimate objects or concepts, or do not see them in people. Rather, their languages have grammatical features that English lacks. True, words like *he*, *she*, and *it* do reflect the sex of their antecedent (except for a few oddities, like referring to a ship as "she"). But their equivalents in French and German refer not to sex but to gender, which is an entirely linguistic, and therefore arbitrary, matter. No French speaker regards a table as having any feminine properties other than grammatical ones.

In more remote languages the differences are still greater. When a Chinese speaker counts items, he or she puts a "measure word" between the number and the item: "one [measure word] book," and so forth. There is nothing quite like this in English, when we arrange numbers in order we signify that we are ordering rather than counting by inserting expressions like number, No.: "We're number 1" "Love Potion No. Nine" and the like. But our practice is invariable, while the Chinese measure word is not; it varies according to the thing being counted. The most common one is *ge*, "one ge person." But for flat objects it is *zhang*, "one *zhang* table"; *and for other kinds of objects there are many other measure words.* Sometimes it is far from obvious what the objects have in common that makes them take a common measure word: the measure word *ba* is used for both chairs and umbrellas!

This all sounds formidably difficult, but only to us—not to the Chinese. The Mandarin variety of Chinese is the native language of over half a billion people in the world even today, and they all master their language at the same rate and by the same age as English speakers do. No language, no matter how strange and difficult it may seem to outsiders, is too hard for its native speakers to master. All languages are systematic, which makes their complexities intelligible to their native speakers, but each system is arbitrary in its own way, which makes it something of a closed book to others.

Equally, no language is especially "simple," if by that word we mean lacking complexity in its phonological and grammatical systems. More likely, people who speak of simplicity in language have a restricted vocabulary in mind. But even this judgment needs to be well-informed if it is to be at all valid. Of course, some languages have larger vocabularies than others, English may comprise half a million words, depending on your manner of counting, while Chinese language would probably have a markedly no more than 10,000 vocabulary. But that vocabulary might be more subtle than English in those areas of thought and experience vital to its users. For example, Eskimos have many different words for different kinds of snow. Moreover, the tribal vocabulary could rapidly expand to deal with new needs as they come along, by borrowing or creating new words. Borrowing, indeed, is one of the most important ways that the English vocabulary has grown to such size. (And, of course, no individual speaker of English has all its half-million words at his or her disposal.)

So the equation of language with culture, one we tend to make, has two possibilities of misleading us. First, we are likely to judge another culture as "simple" because we do not understand it or even know about it; cultural anthropologists would quickly remedy that error for us. Second, we are likely to think that a "primitive" culture has a primitive language. Yet such languages, we now know, seem forbiddingly complex to outsiders who try to learn them.

Above attitudes are forms of ethnocentricity, either because they are too primitive or they are decadent. Language is very fertile ground for ethnocentricity. We are quick to judge even small differences from our own variety of English as "wrong," either laughable or disgustingly. When another people's language is different in more than just small ways, we are inclined to doubt the native intelligence of those who use it, its adequacy for serious purposes, or both.

A more enlightened and indeed more realistic view is the opposite of ethnocentricity. It often goes by the name of "cultural relativism," but learning the name is not the same thing as adopting the view. Only an objective eye on the facts, and a careful eye on our own attitudes, will raise us above ethnocentricity.

To compare linguistics with the study of other forms of human behavior is instructive, but a still grander comparison comes to mind: In many ways the study of language is like the study of life itself. Languages, like species, come into being, grow, change, are sometimes grafted to each other, and occasionally become extinct; they have their histories and, in the written record, their fossils. The origins of both life and language, and their processes, are mysteries that can be penetrated (if at all) by reasoning from incomplete and perhaps ultimately inadequate evidence. And linguists, the scientists of language, study language and its environment with a biologist's care and intensity in order to approach an understanding of the nature of language itself—the most characteristic attribute of all humanity.

Words and Expressions

intricate ['intrikit]	adj.	错综复杂的;难理解的;
axiom ['æksiəm]	n.	公理;自明之理;原理;格言
attribute [ə'tribjuːt]	vt.	把…归于某人;认为…属于某人[物]
['ætribjuːt]	n.	(人或物)属性,特征;价值
specify ['spesifai]	vt.	具体指定,详述;提出…的条件,明确说明
	vi.	明确提出,详细说明
endeavor [in'devə]	vt. & vi.	尝试,试图;尽力,竭力
	n.	努力,尽力
condemned [kən'demd]	adj.	被责难的,被认为不当的,受谴责的

		v.	condemn 的过去式和过去分词)
porpoise	['pɔːpəs]	n.	<动>鼠海豚
verify	['verifai]	vt.	核实,证明,判定,验证,证实
chimpanzees	[ˌtʃimpən'ziːz]	n.	黑猩猩(chimpanzee 的名词复数)
Subtlety	['sʌtlti]	n.	精妙,巧妙;敏锐,敏感;细微的差别等
circus	['səːkəs]	n.	马戏,马戏团;马戏表演
cosmologist	[kɒz'mɒlədʒist]	n.	宇宙学家
prolonged	[prə'lɒŋd]	adj.	延长的,拖延的,长时间的,长期的
laboriously	[lə'bɔːriəsli]	adv.	艰苦地;费力地;辛勤地;艰难地
property	['prɒpəti]	n.	特性,属性;财产,地产;所有权
arbitrary	['ɑːbitrəri]	adj.	随意的,任性的,随心所欲的;主观的
numeral	['njuːmərəl]	n.	数词,数字
chow	[tʃau]	n.	原产中国的狗(体壮,有褐或黑色厚毛,舌为深蓝色)
shepherd	['ʃepəd]	n.	牧羊人,羊倌
overlap	[ˌəuvə'læp]	n.	重叠部分;[数]交叠,相交
intelligible	[in'telidʒəbl]	adj.	可理解的,明白易懂的,清楚的
bureaucratese	[bjuərək'reitiːz]	n.	官僚语言
embody	[im'bɒdi]	vt.	表现,象征;包含,收录;使具体化;[军]组编
triangularity	[traiæŋgju'læriti]	n.	成三角形,三角形变
duality	[djuː'æləti]	n.	二重性,二元性,对偶性
discrete	[di'skriːt]	adj.	非连续的,分离的,不相关联的,分立式
analogy	[ə'nælədʒi]	n.	类似,相似;比拟,类比;类推
reflexive	[rifleksiv]	adj.	反射(性)的 <语>(词或词形)反身的
involuntarily	[in'vɒlənˌterəli]	adv.	非自愿地;无意地;偶然地
jerk	[dʒəːk]	vt. & vi.	(使…)猝然一动[颤动]
parentage	['pɛərəntidʒ]	n.	出身,门第;父母亲的身份,渊源;血统
kitten	['kitn]	n.	小猫;小动物

kinesics	[ki'ni:siks]	n.	人体动作学,举止神态学(身体语言学)
alphabet	['ælfəbet]	n.	字母表;字母系统
physiology	[ˌfizi'ɔlədʒi]	n.	生理学;生理机能
makeup	['meikʌp]	n.	构造
adapted	[ə'dæptid]	adj.	适合的
sustain	[sə'stein]	vt.	维持,支撑,支持
intricate	['intrikət]	adj.	错综复杂的;难理解的;曲折;复杂精细的
motor system	['məutə'sistəm]	n.	运动系统
virtuoso	[ˌvə:tʃu'əusəu]	n.	艺术大师,演奏能手;艺术名家;学者
boardroom	['bɔ:dru:m]	n.	会议室
implement	['impliment]	vt.	实施,执行;实现;落实
spinal cord	['spainəl kɔ:d]	n.	脊髓
peripheral	[pə'rifərəl]	adj.	末梢区域的外围神经系统
complementary	[ˌkɔmpli'mentri]	adj.	互补的;补充的
arcuate fasciculus	['a:kjuit fə'sikjuləs]		弓状纤维束
inferential	[ˌinfə'renʃəl]	adj.	可以推论的,据推理得出的
acoustic	[ə'ku:stik]	adj.	听觉的;声学的;音响的
appendage	[ə'pendidʒ]	n.	附属物;依附的人;<生>附属器官
membrane	['membrein]	n.	(动植物体内的)薄膜;隔膜;膜状物
bass	[beis]	adj.	低音的
oval	['əuvəl]	adj.	椭圆形的;卵形的
		n.	椭圆形;椭圆运动场(等);椭圆
cavity	['kæviti]	n.	腔,洞;蛀牙洞 空洞
skull	[skʌl]	n.	颅骨,头盖骨
homo	['hɔməu]	n.	<拉>人,人类(学名)
genus Homo			人属
sapiens	['seipiənz]	adj.	<拉>(类似)现代人的
genus	['dʒi:nəs]	n.	(动植物的)属;类;种;型
override	[ˌəuvə'raid]	vt.	推翻,无视;践踏;优先于;覆盖

vocal	['vəukl]	adj.	声音的,嗓音的;有声音的
mastery	['mæstəri]	n.	精通,熟练;统治,控制,优势
crop up			(石头)裸露各处;突然发生;犯错误
ancestry	['ænsestri]	n.	祖先;世家,名门;[生]系谱;血统
virtuosity	[,və:tʃu'ɔsəti]	n.	精湛技艺;古董收藏家
disregard	[,disri'ga:rd]	vt.	不顾;不理会;漠视,忽视,蔑视,轻视
inanimate	[in'ænəmit]	adj.	无生命的,无生气的,无精打采的,单调的
antecedent	[,ænti'si:dnt]	n.	经历;<数>(比例)前项;<逻>前件;祖先
gender	['dʒendə(r)]	n.	语言(词)阴中阳性;性别
love potion		n.	春药
formidably	[fə'midəbli]	adv.	可怕地,难对付地,强大地
a closed book			谜,未揭开的秘密,高深莫测的事物
anthropologist	[,ænθrə'pɔlədʒist]	n.	人类学家
forbiddingly		adv.	令人生畏地;严峻地;险恶地
ethnocentricity		n.	民族优越感
decadent	['dekədənt]	adj.	堕落的,颓废的;衰微的
enlightened	[in'laitnd]	adj.	开明的;有知识的;有见识的;
instructive	[in'strʌktiv]	adj.	有益的;教育性的,有启发性的
graft	[gra:ft]	n.	移植;嫁接;渎职;贪污,受贿
penetrated	['penitreitid]	vt.	洞悉,明了,穿透,渗入;潜入

Notes

1. Please refer to Eric H. Lenneberg, *Biological Foundations of Language* (*New York: John Wiley*).

Questions for Discussion and Review

1. What is meant by saying that "language is productive"? Why is "produc-

tivity" one of the most distinctive properties?

2. Another important property of human language is that it is arbitrary. Discuss the several aspects of language characterized by this property.

3. Why is duality of patterning regarded as an important and unique feature of human language?

4. What is "discreteness" of language? Can you explain it in your own words?

5. Review the seven additional "design features" discussed by Bolton. Do they seem to you to be of equal importance? Why or why not? Can you think of any other features not mentioned by Bolton?

6. Why does Bolton insist that "all languages are systematic" and that "no language is especially 'simple'"? In what way is an understanding of these principles important to our understanding of different cultures and their peoples?

Unit 4

The Origin of Language

> The question of how language originated, like the question of how life or world began, has captured the imagination and commanded the attention of generations. A lot of theories sound interesting, fanciful but are speculative and groundless. In this excerpt from his Aspects of language, Bolinger, the professor Emeritus at Harvard University, tries to draw together evidence from a number of fields to explain possible origins of human language and its evolution over the past half-million years. "Language was not stumbled on, invented, or given by the Gods. It grows with us, almost as intimately as our arms and legs."

Not many years ago this chapter would have been forbidden ground. "Speculation about the prehistoric beginnings of language is not a respectable activity." Origins were not to be talked about because they could not be investigated, only guessed at. Known linguistic traces go back to about 5000 B. C. , but beyond that nothing is recoverable.

Anthropologists and anatomists have not been so easily put off and since about 1965 there has been a growing interest in the evidence—mostly from outside linguistics—for at least some notion of when human beings started to speak and how they did it. various lines of investigation have converged: observations of young children, whose speech has primitive beginning very like what can be supposed for early language and may "recapitulate" it; measurement of human skeletons to see whether they had the prerequisites for speaking; investigations of non-speech systems, such as the Sign language of the deaf, to see how necessary actual speech is to the process; the study of natural animal communication; experiments in teaching apes to communicate, to ascertain whether language requires fully developed human intelligence; and of course the excavation of archeological remains, especially of tools, which have their own peculiar relationship to language.

It almost goes without saying that the primordial language, meaning the first that was ever spoken, can never be reconstructed in the way we can reconstruct much of indo-European from bits and pieces of its descendants. According to one view, the rate of change observed in all living languages, if it operated in the past as it does today, would have wiped out traces of any language spoken 30,000 years ago; any similarities that might be found would then simply be due to chance. But there is reason to believe that language much as we know it today existed thousands of years earlier. So the best we can do is to make some informed guesses about what early stages of language were like, not what they were in fact.

What most older theories about origins had in common was a tacit belief in the existence of language as something separate from people. For the Bible it lies at the root of creation: *In the beginning was the word.* For the eighteenth-century philosophers it was invented, man was there beforehand, accoutered with all the powers that he has today except for speech. For speculative linguists it was discovered, in a kind of how-to way: you can bark like a dog to represent "dog," go ding-dong to represent "bell," and say ta-ta on leaving a friend, "waving" goodbye to him with your tongue. (One still hears echoes of the discovery theory, modified to mean that man discovers what is already in his head.) The insights of the theory of evolution came a century late: it took that long to realize that language was simply part of the development of the human race, inseparable from other physical and mental powers, modifying and being modified by them. All life forms transmit. Some use sounds, others smell, touch, taste, movement, temperature changes, or electrical charges. Their messages maintain social unity, warn off predators, attract mates, point to sources of food, and otherwise help keep the species going. As Darwin made clear a century ago, the facial expressions that back up much of human language are an extension of those used by all the primates. There was probably no quantum break with this past; too much of it is still with us. But there must have emerged a succession of differences, important enough to select for survival only those human beings who possessed them to a higher and higher degree.

The first great barrier in animal communication that had to be surmounted was *fixity of reference*. Most animal messages are connected with just one thing in the real world: a growl is a warning to an enemy; a particular scent is an attraction to a mate; a cluck is a summons to a brood of chicks. A dog does not come to his master and growl to indicate that there is an enemy approaching; the growl is at what stimulates it. But transferred meanings are the rule in human language. Basically one steals a purse or a paycheck, but one may also steal a base or steal the limelight. Form and meaning are ditched from each other and to some extent go their own ways. It would be unfair to other species to say that no such detachment is to be found in the non-human world. There are birds that appear to vary their song in ways that are not instinctively predetermined, and neighboring species of fireflies do not all use the same courtship flashes. But this degree of freedom is rare. It is also rare in the human young up to a certain age. But children soon learn that orange not only means a particular piece of fruit regardless of what position or condition of lighting it is seen in, but can also be used of an entire class of similar objects.

The second barrier that had to be surmounted was holophrasis, the emitting of just one independent signaling unit at a time. Animal communication appears to lack syntax. Without it, propositional language is impossible—one cannot say anything about anything, but is limited to command-like or exclamation-like utterances, and those in turn are limited to the here and now: the tribe can be warned of an approaching danger but not reminded of what precautions were taken at the last encounter. Again, human infants lack syntax up to about the age of two, but eventually get the hang of putting words together.

There were further barriers, but these two had to be leveled first, and we must ask why all the life forms on earth had to wait millions of years for Homo to do it. Was it because a certain level of intelligence had to be developed first? But how could that be, when intelligence seems so dependent on language? Language could hardly have been a precondition for language.

The answer seems to be that something a step lower than human intelligence is enough to surmount the first two barriers. Here we can learn from the experiments with two famous chimpanzees, Washoe and Sarah. Washoe learned the American Sign language to the point of transmitting not only one-word messages but messages using combinations as well, and the signs that she used acquired the same flexibility that human words have. For example, she learned the sign for "more" as a way to get her trainers to keep tickling her, but then transferred the sign to a game of being pushed in a laundry cart, and afterwards ex-  tended it spontaneously to swinging by the arms and eventually to ask for the continuation of any activity. Examples of two-word sign-sentences that she invented were "Open food-drink" for "Open the refrigerator"—her trainers had regularly used "cold box" for "refrigerator"—and "Open flower" for a request to be led through a gate to a flower garden. As for Sarah, she used visual signs also, but in place of movements of the hands and fingers she was given plastic tokens of various shapes and colors. These she learned to a point where she was able to interpret fairly complex commands, amounting, for example, to compound sentences. In one experiment, after being taught to respond correctly to "Sarah insert apple pail," "Sarah insert banana pail," "Sarah insert apple dish," and "Sarah insert banana dish," she was confronted with "Sarah insert apple pail Sarah insert banana dish," which she duly obeyed; and finally the redundant words were omitted so that the command read "Sarah insert apple pail banana dish," and again she responded correctly. To do so she had to recognize that "banana" went with "dish" and not with "pail," despite the fact that it was between the names of the two receptacles. Sarah's accomplishments have been replicated by Lana, who punches her symbols on a computer and manages to keep in the news with her expanding vocabulary. Whether these chimps will be able to go on to higher things no one knows—one critic feels that Sarah "is near the limit of her abilities, even with clever stage managing"—but both Sarah and Washoe have matched the language ability of a four-year-old child and have proved that creatures other than humans had the intelligence to transfer meaning and to create syntax.

In a more recent experiment, the trainers of Washoe have worked with a very young chimpanzee, Moja, who at six months already had command of fifteen signs and was putting two of them together to say *Gimme more*! This precociousness suggests one reason for the lower development of chimpanzees: a complete language system is so complex that it requires a prolonged period of plasticity to acquire it. Chimpanzees grow up too fast.

The fact that chimpanzees had the intelligence to create syntax does not of course prove that they did create it, or even that what the present generation of clever chimpanzees have learned can be transmitted from ape to ape; perhaps only the determination and diligence of human trainers could bring it off. But the brain capacity is there.

The third barrier in communication was lack of *metalanguage*. Until elements of language were introduced that referred to language, it was not possible to turn syntax inward and enable it to build on itself. Take as simple an element as a relative pronoun. In a sentence such as Do you know the man who wrote this? the word *who* facilitates the concatenation of two sentences of which one defines an element in the other. It is not necessary to have a word for this—some languages may do it by position, as *Do you know the wrote-this man?* —but whether function word or function order, something has been devised to warn the hearer that one of the two sentences is not a statement in itself but part of a larger proposition. Syntax would be sterile if these inward-pointing elements—pronouns, conjunctions, prepositions, inflections of verbs, sentence adverbs—had not been added to the speaker's repertory.

Looking back at the three barriers we can see that the surmounting of each one was a further gain in recursiveness, in the power to limit the units at one level of language sufficiently to make them manageable and at the same time, at another level, to use them over and over in building larger units. The breakaway from holophrasis made words available not just for single referents but for classes of them. The attainment of syntax made words usable not just one at a time but in combination. The arrival of metalanguage redoubled syntax. Nor was all this movement in one direction. The pressure toward reusability pushed down as well as up. Hearers can identify countless numbers of words not by following any particular guidelines but by recognizing their general phonic outlines, just as they can recognize faces by general impressions without digitizing them. But we don't have to produce faces and we do have to produce words, and we are not all artists. So out of sheer necessity, as the load grew heavier and heavier, there developed a system of phonological points. The worst amateur, no good at freehand speaking, could acquire a set of these through practice and make a good enough imitation of an artistically spoken word to get by.

The upward barriers must have been at least partially leveled before the downward one was breached. This raises the most hotly debated point of all: Was human language originally spoken? There are reasons to believe that it could not have been, but that it would never have developed as far as it has if it had not become so.

1. Preadaptation

We cannot suppose that any (except perhaps the last) of the long series of evolutionary steps that led to language was actually aimed at what it eventually produced. Rather, like the swimming bladder that preceded the lung, it was a mechanism that was on hand and could be adapted to a new use. The ancestors of the apes that first took to the trees to escape predators had to have the forelimb structure that made tree-climbing possible. That in turn created a selective pressure for still more specialized use of those limbs, and structures properly called "arms" were the result. With the descent from the trees, the specialization of arms from legs led in turn to the possibility of standing and walking erect. This made it useful to modify the position of the head, which had to be moved forward to direct the eyes properly. Some progress must already have been made in that direction during the tree-climbing years, since clinging to a vertical trunk to some extent forces an erect posture. As this posture became the normal one, the larynx was pushed down, changing the configuration of the vocal tract and providing for a wider range of vocal sounds.

The "origin" of language from these adaptive changes covers millions of years. But of course it is the later phases that interest us most, for it is only with the use of the apparatus to carry messages that language as we know it can be said to begin.

No other primate uses the vocal organs to communicate anything but rudimentary warnings and emotive cries. That is why earlier attempts to teach apes to communicate with human beings were failures, and why Washoe was taught a sign language. The normal channel with primates is the visual one. They cannot speak because their vocal mechanism does not permit it: they do not move their tongues during a cry, and the sounds that their larynx produces are mostly *aperiodic*— that is, the sounds are not melodic and cannot be modulated for fine pitch contrasts.

Earlier human forms undoubtedly could manage better than a modern chimpanzee, but they too lacked the physical equipment to match the range of modern man. Reconstruction of skulls and vocal musculature reveals that Neanderthal man could not utter the three most stable vowels, [a], [i], and [u]; it would have been harder for him than for us to create sharply differentiated vowel sounds. He could have produced dental and labial consonants ([d b s z v f]), but may not have been able to make a contrast between nasal and non-nasal. Of course even with just two audible contrasts, a phoneme-like code is possible, as any digital computer will tell you—apes could talk if they knew how. But the fewer the contrasts that can be made, the more laborious the coding becomes, so primates would have required greater intellectual powers, not weaker ones. We can conclude that while the lower range of sounds possible for Neanderthal would not have precluded his talking, they required too high a grade of intelligence for him to do much of it. The drive toward vocal expression had begun, but still had a long way to go.

Meanwhile, as a basis for what was to become spoken language, a fairly elaborate system of gesture must have been in use. Culture was already too far advanced at a time when Homo still did not have the power of speech for there not to have been some way of handing skills down from one generation to the next— literally handing, for tool using was the most important skill and is best taught by demonstration, just as it is today. Washoe's trainers found that "a particularly effective and convenient method of shaping [the signs] consisted of holding Washoe's hands, forming them into a configuration, and putting them through the movements of a sign." There is no great distance between signing how to use a tool and signing other meanings; manual shaping is the easiest way to do it. It would be easier to teach children to speak if we could reach into their mouths and mold their tongues. Even now it is the sounds that children can see that they learn most easily (such as the labials [b p m]) and the most difficult widely used sound is one that is both invisible and involves an unusual percussion movement which might be helped if it could be seen and manually shaped: the tapped or trilled r sound.

Could a gesture language have become as expressive a medium of communication as spoken language? Today, among the congenitally deaf, it is very nearly so.

The America Sign Language has its own scheme of arbitrary units, roughly corresponding to a syllabary in spoken language, which is to say that it is not merely a form of pantomime. The signs doubtless are somewhat more iconic than spoken syllables, but memory tests suggest that the deaf have a store of visual and kinesthetic impressions of components of their signs rather like those stored up by the non-deaf of their phonemes. And sign language is akin to speech in other ways. Though it was invented, it is now handed down by tradition and is undergoing the same kinds of changes as natural language. It is somewhat like a pidgin language, first imposed by necessity and then going native, becoming "creolized." An example of internal change is the word for "sweetheart," which half a century ago was formed by bringing the hands together at the edge of the little fingers and cupping the heart, but now is formed with all the fingers in contact except the thumb, more in keeping with other signs in the system. The word has been "leveled," as when a child alters *broke to breaked* to conform to other *-ed* words.

Whether sign language as elaborate as this could have been developed without being preceded by spoken language, or whether deaf people even today could survive if they were not sheltered by a society that mostly uses talk, are unanswered questions that may raise some doubts. But then it is not necessary to suppose that primitive sign language were quite so advanced. It is enough to assume that there was gesture, that those skilled in using it had a better chance to survive, and therefore that any improvement in it would have been reinforced by natural selection. This set the intellectual stage for the transfer to speech, while the physiological stage was probably set by factors already mentioned—man's increasingly erect posture with its lowering of the larynx and the "bent tube" enlargement of the vocal tract. At the same time that the hands were no longer needed for locomotion and were free for tool using and for gestures, the vocal mechanism was as if by accident being prepared to take over. And skill with tools was making it easier to lay up a supply of food so as to restrict the need to chew to a few relatively brief periods, which in turn freed the mouth and tongue for more verbal play.

These advances were reflected in a more complex social organization. If skill with tools is to be transmitted continuously it requires more than the tradition of a single family. As social interdependence increased, dependence on instincts was lessened, and this made for greater resilience in adapting to the environment. An ice age would not wipe out a race that could keep warm by clothing itself rather than having to pass through the tedious evolutionary stages that might develop more "natural" modes of protection such as body hair. Instinctive behavior receded farther and farther into the background, and what we call intelligence superseded it. Mere input-output sensory processing was no longer enough; as early as the beginning of tree-living, it was necessary to have a sharp visual pattern-identifying ability. From this there arose a conceptual level of reality in which human beings acquired a "holistic awareness of entities generically categorized in terms of both their physical attributes and capacity for action." Such categories were ripe for naming. As they were shaped in contact with things, gestural naming was a natural first step.

What makes it seem the more likely that the skills of tool using and language were tied together in man's prehistory is that the brain itself houses them in the same general region. The human brain is lateralized, with functions calling for analysis—tool using, language, symbolic behavior in general—largely confined to one side (the left, for most people, which controls the right side of the body), and space perception, environmental sensitivities, and holistic appreciation confined to the other or more evenly distributed between the two. One can see certain analogies in this kind of brain specialization between the special ways that tools are used and sentences are constructed. With tools, the left hand develops a holding grip while the right develops various precision grips—it "does something to" what is held in the left hand. A propositional sentence contains a topic (usually the grammatical subject) and a comment (usually the grammatical predicate), which does something to or tells something about the topic. In a discourse; the topic is often "held over"—our imagery suggests the analogy with handedness.

Whereas language and tool using are related in the brain, language and primitive cries are not. In man an electrical stimulus on the cortex—the region of highest organization—will cause vocalization; in animals the stimulus generally has to be applied below the cortex. This makes it highly unlikely that there was any direct transition from emotional noises to prepositional language.

If tool using enabled gesture to become practical as well as emotional and provided the push for a systematic gestural language (probably a little more subtle than Washoe's accomplishments with sign language thus far), it also helped to pave the way for the transfer to speech. One can explain the use of a tool through gesture only up to a point; beyond that, especially as tools themselves become more complex, using the hands for explanation interferes with using them for manipulation. Even an accidental sound might have been seized upon under those circumstances, but there were undoubtedly already many that were not accidental, such as vocal signs of approval or disapproval, warnings, persuasions—a great part, probably, of what still constitutes the emotive part of language—not to mention signals used in hunting, where concealment (and hence invisibility) would have inhibited the use of gesture. Besides, even now there remains a good deal of gesture in the use of sound, especially intonation—we can often predict a facial expression if we hear a speech melody. So sound and gesture were already overlapping, and the advantages of sound would have reinforced its use: no interference with other activities of the hands and arms, the possibility of communication out of sight of one's hearer and in darkness, and increased speed—even with a sign language designed as efficiently as possible, the deaf today are held to a comparatively slow rate.

Here is how Morris Swadesh describes the possible origin of one vocalized meaning that must have overlapped most of the long period when gesture was the prevailing mode of communication:

The use of nasal phonemes in the negative in so many languages of the world must in some way be related to the prevailing nasal character of the grunt. In English, the vocable of denial is almost always nasal; but it can vary from a nasalized vowel to any of the three nasal consonants: ā! ā,ē! ē,ō,m! m,n! ŋ! ŋ··· why is nasality so common? Surely because it results from the relaxation of the velum···the most usual position of the velum is down, and the most relaxed form of grunt is nasal. The prevalence of nasals in the negative···may therefore be due to the fact that they are based on grunts.

···

Simple nasality expressed relaxation and contentment. Joined with laryngeal constriction, it signified rather displeasure of frustration.

Once the voice had assumed the major burden of communication, the subsequent refinement of language was largely a matter of cognitive growth which, like a liberated slave, demanded more and more of the freedom that an ever more finely tuned vocal made possible. The change to speech was not merely a recoding, vocally, of units already present in gesture. Vast new possibilities were engendered. The advantages of skilled sound-making redounded on the physical mechanism, which was steadily adapted to language and specialized away from digestion. Linguists of the extreme "language is but a cultural artifact" persuasion have argued that speech is merely an overlaid function, making an artificial use of organs that were designed by nature for the intake of food. But the human speech and digestive organs have developed traits in their later evolution that are not advantageous for eating. The shift of the larynx that made for better sound production "has the disadvantage of greatly increasing the chances of choking to death when a swallowed object gets caught in the pharynx··· The only function for which the adult human vocal tract is better suited is speech." Human beings have evolved as speakers. In the process, the many dimensions in which the sound wave could be modified multiplied the number of distinguishable sounds that could be transmitted, and an ever more agile tongue increased the speed of transmission, thus placing a heavier burden on memory and a higher premium on cognition. This had a drastic effect on the linguistic units themselves, which were able to become increasingly condensed and automatic. In leaping from idea to idea we can no more give conscious thought to speech movements than we can pay attention to our feet when we walk or dance. Such watchfulness may be necessary for learning but is a hindrance in performing.

We can assume that the first vocal units were primarily consonantal—the instability of early vowels has already been noted—and independently meaningful. While the idea of a language with words in which vowels have little or no function except as a transition from one consonant to the next or for affective connotations may seem strange, the fact is that such languages still exist—for example, Kabardian in the northern Caucasus and Bella Coola in coastal British Columbia. As the main burden was on single consonants, there were probably more consonantal contrasts than a language would show today. But as more and more meanings had to be expressed, it became increasingly difficult to add more consonants, and some other device was needed. That of doubling consonants (to express plurality or repetition) had probably already been in use, and could most readily be extended to other meanings; and the splitting off of transitional vowels would have occurred also, making vowels distinctive within syllables. Sentence-like combinations of consonant-words would also have provided the raw material for more complex words. In any case, there was probably a stage in which words were made up of two consonants with or without a vowel contrast. For the first time, distinctive units were on the way to becoming meaningless, but they were probably larger than the units that we regard as distinctive today—roughly, syllables rather than phonemes. As long as the total stock of words was not too large, it was possible for speakers to remember how to produce them as wholes. Words were mostly of one syllable and were to a large degree "transparent"—that is, the meanings of their component sounds were still partially preserved. But the syllable was now ready to fade out its meaning and become a mere building block. How this can happen is illustrated with the suffix -let in English, which is transparently "small" in rivulet, but in bracelet and bullet merely serves to flesh out the words—the connection with brace and ball has been lost. The fact that the earliest forms of sound-writing (which represents words by their sounds rather than by pictures of their meaning) were syllabic makes it clear that the syllable was the arbitrary unit of which speakers had become most conscious, though by that time the phonemes were there in latent form.

The phonemic stage probably came as a result of the increasing number of syllabic units as vocabularies grew larger and larger. The problem would not have been with recognizing words; our memories are capable of storing vast quantities of images configurationally, and we recognize faces and voices with no difficulty, and probably process words the same way when we hear them. Saying them so that someone else can recognize them is a different matter. Word-speaking and syllable-speaking are like freehand drawing. When there are not too many pictures to draw, one can store the instruction in an informal way: "put that curl a little over to the left and slant the eye down a bit but not too much." The difficulty of making even two hundred pictures in this way is obvious, and the number of words in all living languages runs to thousands. So it happened that formerly meaningful sounds were downgraded to a set of phonetic instructions. Instead of "put this curl a little to the left," children learn to "hit the /t/ phoneme in second position in this word"—like following a diagram numbered to guide the amateur's hand in drawing the picture of a face.

The beauty of this is that speakers can not only draw words as effectively as if they were real word-artists, but can exploit to the best advantage the limited range of their vocal apparatus. Only a comparatively small number of phonemes

47

are needed; it is not necessary to crowd them; they can be made sharply distinct. Instead of five or six position along the palatal ridge, two or three suffice, a good distance apart. Instead of a dozen tongue heights, three to five are enough. Even though you miss the target a bit, it is still easy to make each unit distinct from the rest.

This brought language to the stage where all the components it has today were either realized or on the verge. Complete realization was again adaptive. It meant the power to outwit natural enemies. It made possible a new kind of teamwork. Imagine it in operation in driving off or capturing a predator. Complex messages could be transmitted instantly to give precise instructions and keep members of the team informed of directions, locations, and movements. In competition with another tribe it spelt the difference between survival and extinction. Languages—if they and their speakers were to exist at all—had to reach the same level of sophistication in rather short order. For this reason it is not necessary to imagine that all human beings at one time in the past spoke a single language. All that was required was that whatever languages there were should become equally efficient.

2. Variation

The preceding stages can be thought of as the evolution toward language. Subsequent developments are the evolution of language. There is undoubtedly movement in some direction, as there always is; but we are too close to it, and seem rather to be drifting in circles. It is a commonplace among most linguists, for example, that "no 20th-century language is any more advanced than ancient Greek." If we knew what the outcome of the next ten thousand years of evolution was to be, we could measure the factors existing now that are leading to it; but in our ignorance we are unable to mark the signals that are the signs of progress. In any case, it is as if for the whole of recorded history, and no one knows how many hundreds of years before that, languages have merely reshuffled the game-pieces that they inherited from an unread past in an endless array of assortments without any difference in the rules of the game.

What we are most conscious of is that endless array, and the apparent change-without-progress kind of evolution whereby a given language at one stage is converted into a different language at a later stage. What causes it? There are millions of little causes, but the cornerstone of the Tower of Babel was that "ultimate" achievement of an almost meaningless layer, first of syllables and then of phonemes. It increased the symbolizing power of language geometrically, but sacrificed nearly every remnant of mutual intelligibility between dialects from tribe to tribe. Concept and symbol were "freed from each other to the extent that change could modify either one without affecting the other." Change can be fairly rapid, and when groups of speakers are separated for any length of time they end up by not understanding one another, especially if they come into steady contact with speakers of some other language and there is extensive cultural and linguistic intermixture. There is also a certain tendency toward private language that aggravates this—not all speakers at all times want their language to be understood by others. In Ojibwa there is a fairly elaborate etiquette on how not to inform a questioner if his questions are unnecessary or impertinent or if the proper ground has not been laid for them by certain initial formalities. In the Second World War Japanese cryptographers were baffled by an apparently unbreakable code being used in American military communications. They were not aware that Navaho Indians had been recruited to transmit the messages, which were undecipherable because they had no structural relationship with English. But mostly, unintelligibility is not intentional and results instead from the fact that sounds can evolve on their own without destroying meaning, so long as meaning is free to marry itself to any sound. This freedom is not quite total, but it is nearly enough so to have created an enormous variety of languages.

3. Monogenesis or Polygenesis?

Some hints have already been dropped about whether all of today's languages are traceable to a common ancestor. The legend of a single primordial language persists like the legend of Adam and Eve, the single pair from whom all other human beings have sprung. The fact that we can trace many divergent languages today back to one ancestor—Russian and Czech, for example, to common Slavic—suggests that if one were to go back far enough all lines would converge. In the same way, allowing for intermarriage, only two individuals at the outset could account for the whole human family tree. Yet if all life forms as we know them have evolved from earlier forms, it would be strange if at every evolutionary juncture just one pair, conveniently of opposite sexes, should have come into existence to serve as the ancestors of all later forms of that species. It would be just as strange if there had been a moment at which language as the possession of just one society which thereafter transmitted it to its descendants and perhaps to imitators in other societies. Far more likely would be the presence of about as many languages as there were societies, rivaling one another in efficiency and improving through competition. The similarities that are often cited as indications of common source are as easily explained by the fact that human beings are built alike, live out their lives in the same world, and confront the same kinds of problems, and are often in close contact with one another, so that one society borrows from another even though their cultures may earlier have been quite dissimilar. All languages are apt to avoid an initial [fs] bluster because an [f] in that position would scarcely be audible, unlike the media [fs] in *offshoot*. This is in the nature of the speech and hearing mechanism. All language will have nouns because nature confronts us with recurring entities that have to be named and dealt with. And when Chinese cooking is imitated, *chow mein* and *too yong* are borrowed along with it.

But there is another side to the question of monogenesis. Even granting that no time was there just one language, it would still be possible for all the languages spoken today to have descended from a single ancestor, and it is still more possible, even possible, that today's language have a higher degree of kinship than we have imagined.

The pessimistic statement that the primordial language can never be reconstructed, thus needs to be qualified. We probably can never reconstruct a common ancestor for all languages (if indeed they had one), but it may still be possible to reach far back into prehistory and recapture an early ancestor language. One comparatist who believes that something like this may be possible is Mary LeCron Foster, who sees kinships among Asiatic and New world languages, along with proto-Indo-European, that have never been considered to be related. Her technique is to hypothesize sound-changes that will account for differences in primitive roots, and to test the resulting model for consistency and productivity. As with all theoretical models, there comes a point where the predictive power can hardly be due to chance, and then can we be fairly sure that we have at least a shadow of the truth.

4. Look Ahead

There is no more reason for us to suppose that language has played out its evolution. What will be the key to the next surge—social regimentation? International communication? More and more compact societies? One anthropologist would have it that the "graphic period" already marks an epoch, with language change slowed down by writing, and that we are now in the throes of another, a "telecommunicative period," with rate of change retarded still further as recorded models of older speech are imitated. But styles in writing evolve too, and the Network standard is not apt to affect much besides pronunciation. Whatever will mark the new period is here now, in disguise. Like the Promised One, he is among us and we know him not.

Perhaps his name is Implicitness. If meaningful sounds, meaningful syllables, and meaningful words have been successively downgraded to serve as raw material for something yet more intricate, perhaps we are now unconsciously witnessing more and more the downgrading of higher sentences to form implicit elements in macro-sentences. With the word *hopefully* in *hopefully it will be tomorrow*, the speaker enshrines a whole sentence in one word—a sentence which if it were expressed would have to be a main sentence, which the speaker wants to subdue: *It is to be hoped that...* Another example is *reportedly*, as in *Agnew was reportedly incensed at the charge*, where the complete form would be something like *Somebody issued a report; the report said that Agnew was incensed at the charge*. The reduction of sentences has been going on for a long time and may be on the increase, with the result that we shall end by implying more than we say—a fitting development in societies—where people know more and more about their neighbors' affairs.

In any case, the time is past when it was not respectable to speculate on origins or try to forecast the future. Not being required to explain a miracle—the supposed quantum leap from no language to language—we can work comfortably with whatever evolutionary explanations our information will allow. Language was not stumbled on, invented, or given by the gods. It grows with us, almost as intimately as our arms and legs.

Words and Expressions

anatomist	[əˈnætəmist]	n.	解剖学家,剖析者
put off			敷衍;使分心;延期;脱去(衣帽等)
recapitulate	[ˌrikəˈpitʃəˌlet]	vt. & vi.	总结,扼要重述;摘要
prerequisite	[ˌpriːˈrekwəzit]	n.	先决条件,前提,必要条件
		adj.	必须先具备的,必要的;先决条件的
excavation	[ˌekskəˈveiʃn]	n.	挖掘;开凿;开凿的洞穴(或山路等);
archeological	[ˌɑːkiˈrlɒdʒikəl]	adj.	考古学的
primordial	[praiˈmɔːdiəl]	adj.	初生的,初发的,原始的
tacit	[ˈtæsit]	adj.	缄默的,心照不宣的,<律>由法律的效力产生的
accoutered	[əˈkuːtə]	vt.	供以服装,供以军用品,装备
speculative	[ˈspekjələtiv]	adj.	思考的;推理的,揣摩的,推测的
modify	[ˈmɒdifai]	vi.	修改 vt. 改变,减轻,[语]修饰,改变
primate	[ˈpraimeit]	n.	灵长目动物;(英国教会的)大主教
quantum	[ˈkwɒntəm]	n.	[物]量子;定量,总量
surmount	[səˈmaʊnt]	vt.	战胜,克服,登上,攀登
fixity	[ˈfiksiti]	n.	固定性,固定物,不变性
reference	[ˈrefrəns]	n.	参考;提及,参照
growl	[graʊl]	n.	低吼,猞猁声;咆哮
cluck	[klʌk]	n.	(母鸡)咯咯声;<俚>傻瓜,笨蛋
transferred meaning			转义
steal the limelight			抢镜头,最引人注意,出风头
detachment	[diˈtætʃmənt]	n.	分离,分开;分遣
predetermine	[ˌpriːdiˈtəːmin]	vt. & vi.	预先裁定;预先确定;预先决定
holophrasis	[ˌhɒləˈfreisis]	n.	单词句表达(以一个字显示整句意思),句表达
exclamation	[ˌekskləˈmeiʃn]	n.	呼喊,惊叫;感叹语;感叹词

precaution [priˈkɔːʃn]	n.	预防,防备,警惕;预防措施
Homo [ˈhɒməʊ]	n.	人属
redundant [riˈdʌndənt]	adj.	多余的,累赘的
receptacle [riˈseptəkl]	n.	容器,放置物品的地方
punch [pʌntʃ]	vt.	用拳猛击;敲击
chimp [tʃimp]	n.	(非洲)黑猩猩
chimpanzee [ˌtʃimpænˈziː]	n.	黑猩猩,大猩猩
gimme [ˈgimi]	n.	<口>给我(或把它给我)
precociousness [priˈkəʊʃəsnəs]	n.	早熟,早成
plasticity [plæˈstisəti]	n.	柔软性;<生>可塑性(生物体对环境的适应性)
bring off [briŋ ɔf]		使脱离险境;成功完成
metalanguage	n.	语言分析用的语言
concatenation [kənˌkætəˈneiʃn]	n.	互相关联的事
sterile [ˈsterail]	adj.	无效果的,不毛贫瘠的;不生育的;无菌的
inflection [inˈflekʃn]	n.	词尾变化,变音转调,词形变化
repertory [ˈrepətɔːri]	n.	仓库,(知识等的)贮藏,储备
recursiveness [riˈkɜːsivnəs]	n.	递归性;循环性
the attainment [əˈteinmənt]	n.	达到;成就,造诣;学识
reusability [rjuːzəˈbiliti]	n.	可重用性,复用性;可复用性
amateur [ˈæmətər]	n.	业余爱好者;外行,生手
preadaptation [ˈpriːædæpˈteiʃən]	n.	预适应(性);前适应
bladder [ˈblædə(r)]	n.	膀胱;囊状物;水疱,气泡;球胆
precede [priˈsiːd]	vt. & vi.	领先于;先于
descent [diˈsent]	n.	下降;血统;倾斜
posture [ˈpɒstʃə(r)]	n.	姿势;看法;态度;立场
larynx [ˈlæriŋks]	n.	喉,咽喉,喉部
configuration [kənˌfigəˈreiʃn]	n.	布局,构造;配置;[化](分子)组态,排列
vocal tract [ˈvəʊkəl trækt]	n.	声道
apparatus [æpəˈreitəs]	n.	器官,仪器,器械;机构,机关
primate [ˈpraimeit]	n.	灵长目动物;灵长类动物(英国教会的)大主教

英文	音标	词性	中文
rudimentary	[ˌruːdɪˈmentrɪ]	adj.	基本的,初步的;发育不完全的,未成熟的
vocal	[ˈvəʊkl]	adj.	声音的,嗓音的;由嗓音发出或产生的;
melodic	[məˈlɔdɪk]	adj.	有旋律的;调子美妙的
fine pitch	[faɪn pɪtʃ]		精细音调,小螺距,小口距,细距
musculature	[ˈmʌskjələtʃə(r)]	n.	肌肉组织
neanderthal	[nɪˈændətɑːl]	adj.	穴居人的,尼安德特人
differentiated	[dɪfəˈrenʃieɪtɪd]	adj.	已分化的,可区分的,有区别的,差异化的
labial	[ˈleɪbɪəl]	n.	唇音
consonant	[ˈkɒnsənənt]	n.	辅音
nasal	[ˈneɪzl]	n.	鼻音;鼻音字母,鼻骨
phoneme	[ˈfəʊniːm]	n.	音位,音素
preclude	[prɪˈkluːd]	vt.	阻止;排除;妨碍
manual	[ˈmænjuəl]	adj.	用手的;手工的;[法]体力的
percussion	[pəˈkʌʃn]	n.	敲击,碰撞;振动;冲击
trilled	[trɪld]	adj.	颤音的
congenitally	[kənˈdʒenɪtəli]	adv.	天生地,先天地
scheme	[skiːm]	n.	体系;计划;阴谋
corresponding to			相当于…,与…相一致
syllabary	[ˈsɪləbəri]	n.	字音表,音节表
pantomime	[ˈpæntəˌmaɪm]	n.	哑剧;童话剧;手势
iconic	[aɪˈkɒnɪk]	adj.	图标的;图符的;形象的,偶像的
kinesthetic	[ˌkɪnɪsˈθetɪk]	adj.	肌肉运动知觉的
akin	[əˈkɪn]	adj.	同族的;相似的;同源的;关系密切
pidgin	[ˈpɪdʒɪn]	n.	洋泾浜语(不同语种的)混杂语
go native			(移民)接受当地人的(而放弃自己的)风俗习惯
creolize	[ˈkriːəlaɪz]	vt.	使克里奥尔化,使成混合语
in keeping with			和…一致,与…协调
conform to		v.	符合,遵照;顺应;相合
precede	[prɪˈsiːd]	vt. & vi.	先于;领先于
locomotion	[ˌləʊkəˈməʊʃn]	n.	运动;移动;位置移动

lay up	[ri'ziliəns]		贮存;储蓄
resilience	[ri'ziliəns]	n.	弹性;弹力;韧性
supersed	[,su:pə'si:d]	vt.	取代,接替
holistic	[həu'listik]	adj.	全盘的,整体的;功能整体性的
generically	[dʒə'nerikli]	adv.	一般地
attribute	['ætribju:t]	n.	属性,特性,特质,属性
ripe for	[raip fɔ:]		时机成熟,准备就绪
analogies	[ə'nælədʒiz]	n.	类似(analogy 的名词复数);相似;
propositional	[prapə'ziʃənl]	adj.	建议的,提议的,命题的,命题式的
predicate	['predikət]	n.	谓语
discourse	['diskɔ:s]	n.	论述,交谈;正式的讨论
held over			延期;延续;保留;继续上映
analogy	[ə'nælədʒi]	n.	类似,相似;比拟,类比;类推
inhibited	[in'hibit]	v.	抑制;禁止
overlap	[,əuvə'læp]	v.	重叠,部分相同,部分重叠;交叠; 焊瘤
hearer	['hiərə(r)]	n.	听者,听众
subsequent	['sʌbsikwənt]	adj.	后来的;随后的;作为结果而发生的
engender	[in'dʒendə(r)]	vt.	产生;造成;引起
agile	['ædʒail]	adj.	灵巧的;轻快的;灵活的;机敏的
drastic	['dræstik]	adj.	激烈的;猛烈的
caucasus	['kɔ:kəsəs]	n.	高加索山脉,高加索
fade out	[feid aut]		(使)渐隐,渐弱;淡出
transparently	[træns'pærəntli]	adv.	明亮地,显然地
rivulet	['rivjələt]	n.	小河,小溪
bracelet	['breislət]	n.	手镯;手铐
flesh out	[fleʃ aut]		充实,使有血有肉
latent	['leitnt]	adj.	潜在的;潜伏的;休眠的
slant	[slɑ:nt]	vt. & vi.	(使)倾斜
palatal	['pælətl]	adj.	腭的;(指声音)腭音的
outwit	[,aut'wit]	vt.	以智取胜
subsequent	['sʌbsikwənt]	adj.	后来的;随后的
reshuffle	[,ri:'ʃʌfl]	n.	改组(政治组织);重新洗牌
array	[ə'rei]	n.	队列,阵列;数组

assortments	[əˈsɔːtmənts]	n.	分类，各类或同类各种物品混合物
babel	[ˈbeibl]	n.	(圣经)巴别塔；通天塔
geometrically	[ˌdʒiːəˈmetrikli]	adv.	用几何学，几何学上
remnant	[ˈremnənt]	n.	剩余部分，残余物；零料，遗迹
intelligibility	[inˌtelidʒəˈbiləti]	n.	可理解性，可理解的事物
intermixture	[ˌintəˈmikstʃə]	n.	混合，混合物
aggravates	[ˈæɡrəveits]	v.	使恶化；使更严重；
ojibwa			奥吉布瓦人(美洲土著)
etiquette	[ˈetɪket]	n.	礼仪，礼节；规矩
impertinent	[imˈpɜːtinənt]	adj.	无礼的，莽撞的，不切题的，不中肯的
formality	[fɔːˈmæləti]	n.	礼节；拘谨；正式手续
cryptographer	[kripˈtɒɡrəfə]	n.	译解密码者，密码员
baffle	[ˈbæfl]	vt.	使困惑，使迷惑，使受挫折
recruit	[riˈkruːt]	vt.	招聘，征募；雇用
		vi.	征募新兵
undecipherable	[ˈʌndiˈsaifərəbl]	adj.	不可破译的，难解读的
monogenesis	[ˌmɒnəˈdʒenisis]	n.	一元发生说，单性生殖，单源论
polygenesis	[ˌpɒliˈdʒenəsis]	n.	多源发生说
primordial	[praiˈmɔːdiəl]	adj.	初生的，初发的，原始的
czech	[tʃek]	n.	捷克人[语]
at the outset			一开始；当初
juncture	[ˈdʒʌŋktʃə(r)]	n.	时刻，关键时刻；接合点
thereafter	[ˌðeərˈɑːftə(r)]	adv.	此后，在那之后
play out		vi.	停止
surge	[sɜːdʒ]	n.	汹涌；大浪，波涛
regimentation	[ˌredʒimenˈteiʃn]	n.	组织化，系统化
epoch	[ˈepək]	n.	纪元；时期；新时代
throes	[θrəʊz]	n.	剧痛(如分娩时的阵痛)；挣扎
retard	[riˈtɑːd]	vt.	使减速，妨碍，阻止，推迟
		vi.	减慢，受到阻
be apt to		v.	倾向于
implicitness	[imˈplisitnis]	n.	内隐；隐式
subdue	[səbˈdjuː]	vt.	征服；克制；制服

incense ['Insens]   v.   使愤怒;激怒

Questions for Discussion and Review

1. Explain why the first two barriers that had to be overcome in order for human language to develop were fixity of *reference and holophrasis*. Why was lack of *metalanguage* still another barrier that had to be overcome? In what way did the surmounting of each barrier represent "a further gain in recursiveness"?

2. Comment on the statement that originally human language was gestural rather than spoken? What data support this argument?

3. Explain the connections between language and the use of tools. Consider the effect of brain lateralization.

4. According to Bolinger, human language has developed from primarily consonantal units to syllables, and then to phonemes. Trace the course of this development and explain the pressures for and the logic of this succession.

Unit

What Is Linguistics

> In this article, Professor Winfred P. Lehmann, an American linguist, not only succinctly defines linguistics, but also indicates the topics linguists study. He mentions the kinds of problems in the study of linguistics and describes both the practical applications of such study and the variety of interdisciplinary fields that have been developed in recent years.

Linguistics is still such a recent science that a linguist is often asked questions like: What is linguistics? Why study linguistics? What does a linguist do? In answer to the first question the linguist may reply: Linguistics is the scientific study of language. This response may simply lead to the further question: What does it mean to study language scientifically? This article will discuss these questions.

It may seem puzzling to characterize linguistics as a recent science, because people have always concerned themselves with language. In their speculations about language, the ancient Indians saw it as the medium to knowledge. The ancient Hebrews were concerned with the names of such phenomena, ascribing the origin of some names to God, of others to humans, as in the chapters of Genesis: 1:8,10 "And God called the firmament Heaven…the dry land earth"; 2:20 "And Adam gave names to all cattle, and to the fowl of the air, and to every beast of the field." Such concerns with language, found in other early cultures as well, may indeed be characterized as linguistic, but not scientific.

Nonliterate cultures of today also include language in their speculations. The late John Rupert Firth', an energetic and imaginative Yorkshireman who delighted in stimulating others to think about language, enjoyed confronting nonliterate in Africa and India with the statement that their language had no grammar. They would indignantly deny such an assertion, pointing out that they taught their children to follow certain rules and also to avoid certain patterns of language. They too then have linguistic concerns. Besides having firm opinions about their own language, such non-literates may speak several languages, as do many speakers in literate societies.

In at attempt to explain the scientific study of language the linguist may first define language and then give examples of general areas of concern. The definition may go as follows: Language is a system for the communication of meaning through sounds. More precisely, language, viewed as a system, consists of three subsystems or components: one semantic, one syntactic, and one phonological.

The following paragraphs illustrate some problems in semantic, syntactic, and phonological, or sound, patterns that linguists investigate in their study of a language.

For example, they may study verbs that are parallel in their uses to determine specific similarities and differences of meaning, as in the following:

1. She persuaded Frank to sing the aria.
2. She urged Frank to sing the aria.

In indicating encouragement the words *persuade* and *urge* have similar meaning. But they differ considerably in their implications. Every native speaker of English knows these, though the speaker may not be able to formulate them. Sentence 1 implies that Frank sang the aria; Sentence 2, however, carries no such implication: Frank may or may not have sung it. Somewhat similarly the two following sentences also differ in their implications.

3. Frank happened to know the aria.
4. Frank wanted to know the aria.

Sentence 3, though not Sentence 4, implies that Frank knew the aria. Linguists who study meaning may investigate such problems as the expression of implication in a specific language, for example, English, or even more generally in language. Such study is carried out largely with theoretical aims.

But a linguist is also concerned with varying sentence patterns. It is now clear that languages can be classified into two large sets, in accordance with the position of objects with regard to their verbs. English and most of the European languages are verb-object or VO languages; on the other hand, Japanese, Turkish, and many other languages are object-verb or OV languages. In English one says *John saw the dog*, in Japanese *John wa inu o mita*, literally 'John topic marker dog object marker saw.' Consistent VO and OV languages have specific arrangements for other constructions, such as relative clauses, adjectives, and genitives.

In studying syntactic patterns, linguists are also interested in the study and interpretation of such sentence modifications as negation. Thus the negative form of the first two examples would be:

5. She didn't persuade Frank to sing the aria.
6. She didn't urge Frank to sing the aria.

In sentence 5 the negative applies also to sing the aria, whereas in 6 it applies only to urge. The study of sentence patterns involves identifying the "simple sentences" that are the bases for complex sentences like 1 and 2. if one studies sentences in this way, one must assume that the simple sentence underlying the second part of sentence 5 is:

7. Frank didn't sing the aria.

Working out the simplest description of such problems may make up the major concern of a linguist, who might be further characterized as a grammarian or a syntactician.

A linguist may also be interested in the study of speech sounds and their relationship with one another. The last sound of *sing* [ŋ], for example, may be examined because of its patterning. It contrasts directly with the other two nasals (sounds pronounced with the voice going through the nose) in English, as in sing [siŋ] versus sin [sin] and sang [sæŋ] versus Sam [sæm]. But it often stands before k or g, as in *Frank* [fræŋk] and *longer* [læŋgər]. Moreover, the simple adjective *strong*, pronounced with a final [ŋ], has the alternate form with [g] in the comparative stronger [strɔŋgər]. Observing this pattern, some linguists propose that the basic form is found in the comparative and that the [g] is lost or deleted in the simple adjective.

But if a linguist specializes in problems having to do with meaning, with grammar, and with the sounds of language, she or he might be asked about the aims of such study. For what purpose can this information be used? It may be perfectly fine for a theoretician to try to understand, describe, the intricate patterning of language. But do the findings have any further significance? Are they useful in understanding humankind or society?

As linguists have come to know the intricacies of language to a greater extent, their findings have indeed contributed to the understanding of humans and social activities. The investigation of sound patterns, grammatical patterns, and patterns of meaning is of great interest in the study of language acquisition, one of the important concerns in child development. Children learn language in accordance with stages that are gradually becoming better identified. Only in the past few years has it become clear that young infants distinguish between sounds made by humans and other sounds, such as the ringing of a bell. In their first year they experiment with sounds, babbling a great variety of them. At some point, however, they note that sounds have meaning, and then they begin to discriminate—to learn to use sounds as more than an expression of sadness, happiness, or various desires. When they do, they master first a small set of contrasting sounds, such as those in dada and mama. But they add to these very, so that their vowel system includes the i of *pipi*, the u of *pupu*, and so on, and the consonant system comes to include sounds such as p b m, t d n. This expansion proceeds very rapidly, though sounds like r l w may not be mastered for some time. learning of sound agrees with observations of linguists that vowels like [i a u] are somehow more fundamental, or more simple; similarly, the consonants [p b m t d n] are more fundamental than the consonants of *church* or *judge*. Apparently, the behavior of children is in accord with the general observations of linguists concerning sound systems in language.

Sentence patterns—passive sentences rather than active sentences—are learned in similar progression. Children produce sentences like *Jenny saw the dog* earlier than their passive counterparts, such as *The dog was seen by Jenny*. The study of language acquisition then represents an extension of theoretical linguistics to an everyday situation. It interests psychologists and linguists.

Moreover, such study and its findings have special significance for the education of children who may have disabilities, such as deafness. Linguists have become increasingly involved in language disability problems and means for overcoming them, such as the sign systems that have been developed for the communication of the deaf. They have also become involved in exploring the problems of speakers who have speech difficulties, as in stuttering or the partial loss of language through the kind of brain damage that causes aphasia, the impairment or loss of control over language.

As information on aphasia has increased, it has become clear that loss of language control may involve the meaning system, or grammar, or the sounds, or combinations of these three facets of language. An aphasic may use the label *chair* to refer to a table, or may confuse sentence patterns, or may lose control of certain sounds, often those that are acquired last by children. Linguists have much to learn about such problems, in collaboration with other specialists. But even now a great deal has been learned about lateralization, that is, the functioning of the human brain in such a way that speech is controlled generally by the left hemisphere. If this hemisphere is damaged in children, control of speech may be taken over by the right hemisphere. But this is not true of adults.

Concern with such problems has led to interdisciplinary activities and even to new concentrations. The new specialty psycholinguistics, which deals with those areas involving psychology and the study of language, had scarcely become established before concentrated study of the functioning of the brain with regard to language led to the development of neurolinguistics. As yet, few linguists have acquired the complex knowledge needed to work in such intricate areas. But the number is increasing. More and more linguists, especially students entering the field, will apply their findings in attempting to understand the functioning of the brain in the use of language and to solve problems that arise when such functioning is impeded.

Linguists concentrating on such problems are concerned primarily with individual speakers. Other linguists, often called sociolinguists, are investigating the uses of language in society. For it is clear that social groupings are held together by the use of language. In earlier days, especially when communication was restricted as in the Americas several thousands of years ago, limited communication led to the development of many languages spoken by relatively small numbers of speakers. This was also the situation in Europe in the several millennia before our era. Rome, for example, and a small area around it spoke Latin. Other sections of Italy spoke Etruscan, Greek, Oscan, or other languages. Vigorous communities like that of Rome then expanded their territorial control and extended their language to many new speakers, so that in time all of Italy and eventually most of western Europe was one social and political group, the Roman Empire, using Latin for communication. Some sociolinguists study the ways in which such social groups use language, how the language develops numerous varieties known as dialects, and what happens to these dialects in dynamic societies.

But social groups held together by one language or by a dialect need not be as large as the Roman Empire or our modern nations. Clusters of speakers, even in industrial societies, may constitute separate social communities, such as the foreign workers in many European cities at present or Spanish-speaking communities in American cities. Or the language used by a social group may merely be a special variety of the national language, such as soul talk, student slang, or thieves' argot. Such varieties permit groups to limit communication among themselves; in this way the special variety of language produces a feeling of independence. If the limitation becomes too severe, however, social difficulties may result. Such potential difficulties have come to the attention of modern states, for example, India, the new countries of Africa, and China.

Many linguists in these countries are studying the intricacies of communication resulting from the use of many languages. In making such studies they map the various forms of speech by geographical location or by social class. In India, for example, different languages often correlate with different castes.

As such differences become clear, countries may try to establish one language for general use, as the Roman Empire did Latin. India has focused on Hindi; China has selected a form of mandarin Chinese for its "normal language".

Many intricacies of communication in social groups are just becoming apparent. In spite of the widespread use of mass communication by means of modern media, modern industrial communities still maintain different forms of speech and may even be extending them. Besides identifying these, their functions, and their causes, linguists have come to cooperate with other specialists in attempting to solve resulting problems, such as those leading to isolated communities or isolated segments of communities. Contemporary societies attempt to solve many such problems through their school systems.

The study of language has always made up a large part of school curricula, particularly at the elementary and secondary levels. Students are to be taught to speak effectively, to read capably, and to write competently. In American schools the success of these aims can be achieved the more effectively if teachers are informed concerning the varieties of English and the attitudes toward these varieties. Considerable study is now being done on the identification of dialect differences, of attitudes toward them, and of bilingualism. Together with school teachers, linguists are also attempting to identify the processes involved in reading, especially when these involve difficulties such as dyslexia.

In Chinese schools, teachers with the assistance of linguists are teaching Putonghua to children of different native dialects. The linguists identify the characteristics of such dialects; the teachers then devise means to assist the children in shifting from their former language patterns to those of Putonghua. Linguists are also working in China to devise and teach alphabets, both for Putonghua—which generally is written in characters—and for the minority languages of the country. The language problems in China, as in most nations of the world, will require increasing linguistic information. On the basis of such information, countries like India, Indonesia, and many in Africa will establish their selected national languages. Moreover, measures taken to increase literary will absorb the energies of many linguists.

Linguistics admittedly has no magical solutions for the problems cited. But since these problems involve language, there is every hope that increased understanding of language will lead to some remedies.

Linguistics therefore offers many opportunities. Societies cannot function without language, and their speakers cannot achieve full development without adequate control of language. Linguists will work increasingly with other specialists, especially in the behavioral and the biological sciences, to solve social and individual problems. As insights into language increase, linguists will have increasing opportunities and responsibilities to many problems encountered by individuals and societies.

Notes

1. John Rupert Firth: A British linguist and also a sociolinguist, who founded the "London School" of Linguistics. His influential works include *Papers in Linguistics*, and *The Tongues of Men, and Speech*.

2. See the article by R. E. Callary (p. 109) for an explanation of this symbol and those used later in this article.

Words and Expressions

Hebrews	[ˈhiːbruːz]	n.	(圣经)希伯来书;希伯来人(语),以色列语
ascribe	[əˈskraib]	vt.	把…归于,认为…是由于
genesis	[ˈdʒenisis]	n.	＜创世记＞
fowl	[faul]	n.	鸡;家禽,飞禽
indignantly	[inˈdignəntli]	adv.	愤怒地,愤慨地,愤愤不平地
aria	[ˈɑːriə]	n.	咏叹调;唱腔,唱段
genitives	[ˈdʒenitiv]	n.	＜语＞所有格
intricacies	[ˈintrikəsi]	n.	(复数)错综复杂的事物
discriminate	[diˈskrimineit]	vt. & vi.	区分;歧视;区别
impairment	[imˈpeəmənt]	n.	损害,损伤
aphasia	[əˈfeiziə]	n.	失语症
aphasic	[əˈfeizik]	n. & adj.	患失语症者,失语症的
lateralization	[lætərəlaiˈzeiʃən]	n.	(尤指脑部的)偏侧性,偏侧优势,偏利
intricate	[ˈintrikit]	adj.	错综复杂的;难理解的;
imped	[imˈpiːd]	vt.	阻碍;妨碍;阻止
Etruscan			意大利中西部古国伊特鲁里亚
oscan	[ˈɔskən]	n.	欧斯干人
argot	[ˈɑːrgət]	n.	行话,黑话
castes	[ˈkɑːsts]	n.	印度的种姓社会等级
curricula	[kəˈrikjələ]	n.	课程;总课程(curriculum 的名词复数)
dyslexia	[disˈleksiə]	n.	读写困难,读写障碍
alphabet	[ˈælfəbet]	n.	字母表;字母系统;入门,初步

Questions for Discussion and Review

1. Talking into consideration the definition of linguistics given at the beginning of the article by Lehmann, explain the statement towards the end of the article:

"Linguistics then can scarcely be a pursuit restricted to the scientific study of language,. ···" Why does Lehmann say so?

2. According to Lehmann, language consists of three subsystems. What are they? Give an example of your own, for each subsystem, that illustrates a problem that linguists might find interesting.

3. What are some of the practical applications of the work of linguists?

4. What are some of the interdisciplinary subjects involving linguistics? With what topics or areas do they deal?

Unit 6

The Object of Linguistics

> In this excerpt from his famous work *Course in General Linguistics*, Ferdinand de Saussure, a structuralist of the European tradition, defines language in a classic manner—language is "a system of signs and a self-contained whole, a grammatical system that has a potential existence… in the brains of a group of individuals." Saussure draws several distinctions, two of which, the distinction between *langue* and *parole* and the distinction between the signifier and the signified, are touched upon in this excerpt. At the University of Geneva, he taught comparative grammar, Sanskrit and general linguistics(1907—1911). His "influence on the development of linguistics was decisive." (Wade Baskin)

1. Language and Speech

What is language [*langue*]? It is not to be confused with human speech [*parole*], of which it is only a definite part, though certainly an essential one. language is both a social product of the faculty of speech and a collection of necessary conventions that have been adopted by a social body to permit individuals to exercise that faculty. Language is a self-contained whole and a principle of classification, it exists perfectly only within a collectivity. In separating language from speech we are at the same time separating: (1) what is social from what is individual; and (2) what is essential from what is accessory and less accidental.

Language is not a function of the speaker; it is a product that is passively assimilated by the individual. It never requires premeditation, and reflection in only for the purpose of classification. Speech, on the contrary, is an individual act. Within the act, we should distinguish between (1) the combinations by which the speaker uses the language code for expressing his own thought; and (2) the psychophysical mechanism that allows him to exteriorize those combinations.

In Latin, *articulus* means a member, part, or subdivision of a sequence; applied to speech, *articulation* designates either the subdivision of a spoken chain into syllables or the subdivision of the chain of meanings into significant units; we can say that what is natural to mankind is not oral speech but the faculty of constructing a language, i. e. a system of distinct signs corresponding to distinct ideas.

To give language first place in the study of speech, we can advance a argument: the faculty of articulating words—whether it is natural or not—is exercised only with the help of the instrument created by a collectivity and provided for its use; therefore, to say that language gives unity to speech is not fanciful.

These are the characteristics of language:

(1) Language is a well-defined object in the heterogeneous mass of speech facts. It can be localized in the limited segment of the speaking-circuit where an auditory image becomes associated with a concept. It is the social side of speech, outside the individual who can never create nor modify it by himself; it exists only by virtue of a sort of contract signed by the members of a community. Moreover, the individual must always serve an apprenticeship in order to learn the functioning of language; a child assimilates it only gradually. It is such a distinct thing that a man deprived of the use of speaking retains it provided that he understands the vocal signs that he hears.

(2) Language, unlike speaking, is something that we can study separately. Although dead languages are no longer spoken, we can easily assimilate their linguistic organisms. We can dispense with the other elements of speech; indeed, the science of language is possible only if the other elements are excluded.

(3) Whereas speech is heterogeneous, language, as defined, is homogeneous. It is a system of signs in which the only essential thing is the union of meanings and sound-images, and in which both parts of the sign are psychological.

(4) Language is concrete, no less so than speaking; and this is a help in our

study of it. Linguistic signs, though basically psychological, are not abstractions; associations which bear the stamp of collective approval—and which added together constitute language—are realities that have their seat in the brain. Besides, linguistic signs are tangible; it is possible to reduce them to conventional written symbols, whereas it would be impossible to provide detailed photographs of acts of speaking [*actes de parole*]; the pronunciation of even the smallest word represents an infinite number of muscular movements that could be identified and put into graphic form only with great difficulty. In language, on the contrary, there is only the sound-image, and the latter can be translated into a fixed visual image. For if we disregard the vast number of movements necessary for the realization of sound-images in speaking, we see that each sound-image is nothing more than the sum of a limited number of elements or phonemes that can in turn be called up by a corresponding number of written symbols. The very possibility of putting the things that relate to language into graphic form allows dictionaries and grammars to represent it accurately, for language is a storehouse of sound-images, and writing is the tangible form of those images.

2.Place of Language in Human Facsts: Semiology

The foregoing characteristics of language reveal an even more important characteristic. Language, once its boundaries have been marked off within the speech, can be classified among human phenomena, whereas speech cannot.

We have just seen that language is a social institution; but several features set it apart from other political, legal, etc. institutions. We must call in a new type of facts in order to illuminate the special nature of language.

Language is a system of signs that express ideas, and is therefore comparable to a system of writing, the alphabet of deaf-mutes, symbolic rites, polite formulas, military signals, etc. But it is the most important of all these systems.

A science that studies the life of signs within society is conceivable; it would be a part of *social psychology* and consequently of general psychology; I shall call it semiology (from Greek *semeion* "sign"). Semiology would show what constitutes signs, what laws govern them. Since the science does not yet exist, no one can say what it would be; but it has a right to existence, a place staked out in advance. Linguistics is only a part of the general science of semiology; the laws discovered by semiology will be applicable to linguistics, and the latter will circumscribe a well-defined area within the mass of anthropological facts.

To determine the exact place of semiology is the task of the psychologist. The task of the linguist is to find out what makes language a special system within the mass of semiological data. This issue will be taken up again later; here I wish merely to call attention to one thing: if I have succeeded in assigning linguistics a place among the sciences, it is because I have related it to semiology.

Why has semiology not yet been recognized as an independent science with its own object like all the other sciences? Linguists have been going around in circles: language, better than anything else, offers a basis for understanding the semiological problem, but language must, to put it correctly, be studied in itself; heretofore language has almost always been studied in connection with something else, from other viewpoints.

There is first of all the superficial notion of the general public: people see nothing more than a name-giving system in language, thereby prohibiting any research into its true nature.

Then there is the viewpoint of the psychologist, who studies the sign-mechanism in the individual; this is the easiest method, but it does not lead beyond individual execution and does not reach the sign, which is social.

Or even when signs are studied from a social viewpoint, only the traits that attach language to the other social institutions—those that are more or less voluntary—are emphasized; as a result, the goal is by-passed and the specific characteristics of semiological systems in general and of language in particular are completely ignored. For the distinguishing characteristic of the sign—but the one that is least apparent at first sight—is that in some way it always eludes the individual or social will.

In short, the characteristic that distinguishes semiological systems from all other institutions shows up clearly only in language where it manifests itself in the things which are studied least, and the necessity or specific value of a semiological science is therefore not clearly recognized. But to me the language problem is mainly semiological, and all developments derive their significance from that important fact. If we are to discover the true nature of language we must learn what it has in common with all other semiological systems; linguistic forces that seem very important at first glance (e. g., the role of the vocal apparatus) will receive only secondary consideration if they serve only to set language apart from the other systems. This procedure will do more than to clarify the linguistic problem. By studying rites, customs, etc. as signs, I believe that we shall throw new light on the facts and point up the need for including them in a of semiology and explaining them by its laws.

Words and Expressions

signifier	[ˈsignifaiə(r)]	n.	能指；意符；信号物；记号
signified	[ˈsignifaid]	n.	所指，符号义
sanskrit	[ˈsænskrit]	n.	梵文，梵语
definite	[ˈdefinət]	adj.	明确的；一定的；确定的
self-contained	[self kənˈteind]	adj.	独立的；自给自足的
designate	[ˈdezigneit]	vt.	指明，指出；表明，意味着；指定
faculty	[ˈfæklti]	n.	能力，才能
heterogeneous	[ˌhetərəˈdʒi:niəs]	adj.	各种各样的；成分混杂的，异构的，异类的，异质的
concept	[ˈkɒnsept]	n.	概念；思想
dispense with		vt.	摒弃，省掉；摈除
homogeneous	[ˌhɒməˈdʒi:niəs]	adj.	同性质的，同类的；相似的
conventional	[kənˈvenʃənl]	adj.	传统的；依照惯例的；约定的
photograph	[ˈfəutəgrɑ:f]	n.	照片，相片
infinite	[ˈinfənit]	adj.	无限的，无穷的；无数的，许多的
call up			叫醒；给…打电话
represent	[ˌrepriˈzent]	vt.	表现，象征；描绘
tangible	[ˈtændʒəbl]	adj.	实际的，有形的；具体的
semiology	[ˌsi:miˈɒlədʒi]	n.	符号学，记号学
foregoing	[ˈfɔ:gəuiŋ]	adj.	前面提到的，前面的
marked off			划出，划开；区分
illuminate	[iˈlu:mineit]	vt.	照明，阐明，说明
symbolic	[simˈbɒlik]	adj.	象征的，象征性的，符号的
rite	[rait]	n.	仪式，典礼，礼仪
formula	[ˈfɔ:mjələ]	n.	公式，准则；客套话
conceivable	[kənˈsi:vəbl]	adj.	可想到的，可相信的，可能的

stak out   vt.   明确标明清楚界定
circumscribe ['sə:kəmskraib]   vt.   划定…范围;限定;划界限
heretofore   [ˌhiətu'fɔ:(r)]   adv. 在此以前,迄今为止
superficial   [ˌsu:pə'fiʃl]   adj. 表面(上)的;肤浅的
notion   ['nəʊʃn]   n.   观念;见解
thereby   [ˌðeə'bai]   adv. 由此,从而
sign   [sain]   n.   符号

Questions for Discussion and Review

1. Explain de Saussure's definition of language: Language is a system of signs or a self-contained whole, and compare it with a definition you know of another linguist.

2. Discuss the difference between langue and parole.

3. what is the purpose to distinguish langue and parole?

4. What are the characteristics of language, according to de Saussure?

Unit 7

Nature of the Linguistic Sign

> This is another excerpt from *Course in General linguistics*, Ferdinand de Saussure answers the question "what is a linguistic sign, for example, a word?" He continues with the discussion on the relationship between the signifier and the signified. He exemplifies the linguistic sign as "a two-side psychological entity," the combination of a sound-image(the signifier) and a concept(the signified). The linguistic sign, as he defines, also has two characteristics: the arbitrary nature of the sign and the linear nature of the signifier.

1. Sign, Signified, Signfier

Some people regard language, when reduced to its elements, as a naming-process only—a list of words, each corresponding to the thing it names. For example:

This conception is open to criticism at several points. It assumes that ready-made ideas exist before words; it does not tell us whether a name is vocal or psychological in nature (arbor, for instance, can be considered from either viewpoint); finally, it lets us assume that the linking of a name and a

TRee=arbor Horse=Equos

thing is a very simple operation—an assumption that is anything but true. But this rather naive approach can bring us near the truth by showing us that the linguistic unit is a double entity, one formed by the associating of two terms.

We have seen in considering the speaking-circuit that both terms involved in the linguistic sign are psychological and are united in the brain by an associative bond. This point must be emphasized.

The linguistic sign unites, not a thing and a name, but a concept and a sound-image. The latter is not the material sound, a purely physical thing, but the psychological imprint of the sound, the impression that it makes on our senses. The sound-image is sensory, and if I happen to call it "material," it is only in that sense, and by way of opposing it to the other term of the association, the concept, which is generally more abstract.

The psychological character of our sound-images becomes apparent when we observe our own speech. Without moving our lips or tongue, we can talk to ourselves or recite mentally a selection of verse. Because we regard the words of our language as sound-images, we must avoid speaking of the "phonemes" that make up the words. This term, which suggests vocal activity, is applicable to the spoken word only, to the realization of the inner image in discourse. We can avoid that misunderstanding by speaking of the *sounds and syllables* of a word provided we remember that the names refer to the sound-image.

The linguistic sign is then a two-sided psychological entity that can be represented by drawing:

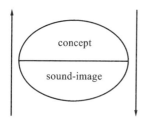

The two elements are intimately united, and each recalls the other. Whether we try to find the meaning of the Latin word arbor or the word that Latin uses to designate the concept "tree," it is clear that only the associations sanctioned by that language appeal to us to conform to reality, and we disregard whatever others might be imagined.

Our definition of the linguistic sign poses an important question of terminology. I call the combination of a concept and a sound-image *sign*, but in current usage the term generally designates only a sound-image, a word, for example (*arbor*, etc.). One tends to forget that arbor is called a sign only because it carries the concept "tree," with the result that the idea of the sensory part implies the idea of the whole.

Ambiguity would disappear if the three notions involved here were designated by three names, each suggesting and opposing the others. I propose to retain the word *sign*[*signe*] to designate the whole and to replace *concept* and *sound-image* respectively by signified [*signifié*] and signifier [*signifiant*]; the last two terms have the advantage of indicating the opposition that separates them from each other and from the whole of which they are parts. As regards *sign*, if I am satisfied with it, this is simply because I do not know of any word to replace it, the ordinary language suggesting no other.

The linguistic sign, as defined, has two primordial characteristics. In enunciating them I am also positing the basic principles of any study of this type.

2. Principle Ⅰ: The Arbitrary Nature of the Sign

The bond between the signifier and the signified is *arbitrary*. Since I mean by sign the whole that results from the associating of the signifier with the signified, I can simply say: *the linguistic sign is arbitrary*.

The idea of "sister" is not linked by any inner relationship to the succession of sound s-ö-r which serves as its signifier in French; that it could be represented equally by just any other sequence is proved by differences among languages and by the very existence of different languages: the signified "ox" has as its signifier b-ö-f on one side of the border and o-k-s (Ochs) on the other.

No one disputes the principle of the arbitrary nature of the sign, but it is often easier to discover a truth than to assign to it its proper place. Principle I dominates all the linguistics of language; its consequences are numberless. It is true that not all of them are equally obvious at first glance; only after many detours does one discover them, and with them the primordial importance of the principle.

One remark in passing: when semiology becomes organized as a science, the

75

question will arise whether or not it properly includes modes of expression based on completely natural signs, such as pantomime. Supposing that the new science welcomes them, its main concern will still be the whole group of systems grounded on the arbitrariness of the sign. In fact, every means of expression used in society is based, in principle, on collective behavior or—what amounts to the same thing—on convention. Polite formulas, for instance, though often imbued with a certain natural expressiveness (as in the case of a Chinese who greets his emperor by bowing down to the ground nine times), are nonetheless fixed by rule; it is this rule and not the intrinsic value of the gestures that obliges one to use them. Signs that are wholly arbitrary realize better than the others the ideal of the semiological process; that is why language, the most complex and universal of all systems of expression, is also the most characteristic; in this sense linguistics can become the master-pattern for all branches of semiology although language is only one particular semiological system.

The word *symbol* has been used to designate the linguistic sign, or more specifically, what is here called the signifier. Principle I in particular weighs against the use of this term. One characteristic of the symbol is that it is never wholly arbitrary; it is not empty, for there is the rudiment of a natural bond between the signifier and the signified. The symbol of justice, a pair of scales, could not be replaced by just any other symbol, such as a chariot.

The word *arbitrary* also calls for comment. The term should not imply that the choice of the signifier is left entirely to the speaker (we shall see below that the individual does not have the power to change a sign in any way once it has become established in the linguistic community); I mean that it is unmotivated, i. e. arbitrary in that it actually has no natural connections with the signified.

In concluding let us consider two objections that might be raised to the establishment of Principle I.

(1) *Onomatopoeia* might be used to prove that the choice of the signifier is not always arbitrary. But onomatopoeic formations are never organic elements of a linguistic system. Besides, their number is much smaller than is generally supposed. Words like French *fouet* 'whip' or *glas* 'knell' may strike certain ears with suggestive sonority, but to see that they have not always had this property we' need only examine their Latin forms (*fouet* is derived from *fāgus* 'beech-tree.' *glas* from classicum 'sound of a trumpet'). The quality of their present

sounds, or rather the quality that is attributed to them, is a fortuitous result of phonetic evolution.

As for authentic onomatopoeic words (e. g. *glug-glug*, *tick-tock*, etc.), not only are they limited in number, but also they are chosen somewhat arbitrarily, for they are only approximate and more or less conventional imitations of certain sounds (cf. English *bow-bow* and French *ouaoua*). In addition, once these words have been introduced into the language, they are to a certain extent subjected to the same evolution—phonetic, morphological, etc. —that other words undergo (cf. *pigeon*, ultimately from Vulgar Latin *pī piō*, derived in turn from an onomatopoeic formation): obvious proof that they lose something of their original character in order to assume that of the linguistic sign in general, which is unmotivated.

(2) *Interjections*, closely related to onomatopoeia, can be attacked on the same and come no closer to refuting our thesis. One is tempted to see in them spontaneous expressions of reality dictated, so to speak, by natural forces. But for most interjections we can show that there is no fixed bond between their signified and their signifier. We need only compare two languages on this point to see how much such expressions differ from one language to the next (e. g. the English equivalent of French aïe! aïe is ouch!). We know, moreover, that many interjections were once words with specific meanings (cf. French diable! 'darn!' mordieu! 'golly!' from mort Dieu 'God's death,' etc.).

Onomatopoeic formations and interjections are of secondary importance, and their symbolic origin is in part open to dispute.

3. Principle II: The Linear Nature of the Signifier

The signifier, being auditory, is unfolded solely in time from which it gets the following characteristics: (a) it represents a span, and (b) the span is measurable in a single dimension; it is a line.

While Principle II is obvious, apparently linguists have always neglected to state it, doubtless because they found it too simple; nevertheless, it is fundamental, and its consequences are incalculable. Its importance equals that of Principle I; the whole mechanism of language depends upon it. In contrast to visual signifiers (nautical, signals, etc.) which can offer simultaneous groupings in several dimensions, auditory signifiers have at their command only the dimension of time. Their elements are presented in succession; they form a chain. This feature becomes readily apparent when they are represented in writing and the spatial line of graphic marks is substituted for succession in time.

Sometimes the linear nature of the signifier is not obvious. When I accent a syllable, for instance, it seems that I am concentrating more than one significant element on the same point. But this is an illusion; the syllable and its accent constitute only one phonational act. There is no duality within the act but only different oppositions to what precedes and what follows.

Words and Expressions

linear	['liniə(r)]	adj.	直线的,线形的
arbitrary	['ɑːbitrəri]	adj.	任意的,任意性的,随心所欲的,主观的
primordial	[prai'mɔːdiəl]	adj.	初生的,初发的,原始的
enunciate	[i'nʌnsi,et]	vi. & vt.	（清晰地）发音；确切地说明,阐明
remark	[ri'mɑːk]	n.	注意,观察；话语；评论,谈论
		vt. & vi.	评论；觉察
semiology	[,siːmi'ɒlədʒI]	n.	记号语言,记号学,符号学
pantomime	['pæntə,maim]	n.	哑剧；童话剧；手势
imbue	[im'bjuː]	vt.	灌输；使感染；浸染
imbued with			充满
rudiment	['ruːdimənt]	n.	基本原理；雏形；萌芽；退化器官
chariot	['tʃæriət]	n.	敞篷双轮马车(古代用于战争)战车
onomatopoeia	['ɒnə,mætə'piːə]	n.	拟声,拟声法
knell	[nel]	n.	丧钟声；某事物结束的象征
beech-tree	[biːtʃ]	n.	[植]山毛榉；山毛榉木材
authentic	[ɔː'θentik]	adj.	真的,真正的；可信的,可靠的
vulgar	['vʌlgə]	adj.	庸俗的,粗俗的；老百姓的；粗野下流的
interjections	[,intə'dʒekʃn]	n.	叹词,感叹词

refute	[ri'fjuːt]	v.	驳斥,驳倒(refute 的现在分词)
spatial	['speiʃl]	adj.	空间的,受空间条件限制的,占大篇幅的
graphic	['græfik]	adj.	图解的,图示的,用文字表示的,形象的
duality	[djuː'æləti]	n.	二元性,二重性
precede	[pri'siːd]	vt. & vi.	在…之前发生或出现,先于;优于

Questions for Discussion and Review

1. Why does de Saussure say that a linguistic sign is a two-sided psychological entity?

2. Exemplify what the signifier is and what the signified is.

3. why does de Saussure describe the linguistic sign itself as arbitrary?

4. explain the linear of the signifier.

Unit 8

Some Basic Concepts in Linguistics

> **In this excerpt from** *Papers in Applied Linguistics* edited by J. P. B. Allen, he starts with the tasks and goals of linguists, explaining macrolinguistics and microlinguistics, and the four guiding principles in scientific analysis; he then goes on to describe the unique features of human language. Finally, he explains the relationship between form and meaning, discussing the formal and meaning-based approaches to language, syntagmatic and paradigmatic relations of linguistic elements.

1. The scientific study of language

Linguistics is a comparative newcomer to the academic scene. Most people think they have a clear idea of what the more traditional subjects of the University curriculum are about. At social gatherings one does not often hear the question "What is chemistry?" or "What is engineering?" A linguist, however, is quite likely to be called upon to expound the basic principles of linguistics several times during the course of an evening. How, then, might a linguist attempt to state, in a few words, the basic characteristics of his subject? He might begin by saying that a linguist is not necessarily someone who is a fluent speaker of foreign languages. Undoubtedly, it is advantageous if a linguist has a practical mastery of one or more languages apart from his own, but this is not essential to his main business, which is the study of language in general. In linguistics, we do not give priority to the language of any particular society; we study the languages of any or all human societies. We study how each language is constructed, how it is used by its speakers, and how it is related to other languages. In addition, we study how it varies from dialect to dialect, and how it changes from one historical period to the next.

Linguistics may be defined as the scientific study of language. This apparently straitforward definition conceals a wide divergence of views about what is meant by "scientific". For our present purposes let us say simply that a "scientific" study is one which is based on the systematic investigation of data, conducted with reference to some general theory of language structure. In the working methods of all linguists, however, theory formation and the study of data have always proceeded side by side. We may say that in linguistics, as in other areas of science, data and theory "stand in a dialectical complementation" and that neither can be profitably studied without the other.

All types of linguistic analysis are based on the assumption that language is structured; i. e. , each utterance, far from being a random series of words, is put together according to some principle, or set of principles, which determines the words that occur and the form and order of the words. In his analysis a linguist is guided by the three major canons of exhaustiveness(adequate treatment of all the relevant material); consistency (the absence of contradiction between different parts of the total statement); and economy, whereby, other things being equal, a shorter statement or analysis employing fewer terms is to be preferred to one that is longer or more involved. Exhaustiveness and economy are interdependent, in that there is no point in setting up a description which is economical but which does not cover all the data. On the other hand, an exhaustive description may not be aesthetically pleasing because of the inevitable exceptions. As a result of applying these criteria, the linguist expects to achieve the greatest possible degree of objectivity in his descriptions. The aim is to present an analysis in such a way that every. part of it can be tested and verified, not only by the author himself, but by anyone else who chooses to refer to it or to make a description of his own based on the same principles.

It is generally useful to distinguish between a broader and a narrower aspect of linguistic study; the terms "macrolinguistics" and "microlinguistics" have been used to describe these two aspects. Macrolinguistics refers to the whole study of language, including such widely diverse fields as psycholinguistics, sociolinguistics, historical linguistics, speech pathology, lexicography, computational linguistics and communication theory. Microlinguistics refers to what may be called the "central core" of language study, the areas of phonology, grammar and semantics. Phonology is the study of the sound patterns of language; grammar is concerned with the form of the words and the manner of their combination in phrases, clauses and sentences; while semantics is concerned with the meaning, or content, of the words and of the larger units into which they combine. In the article we will confine ourselves in the main to a "micro" interpretation of linguistics. Our concern will be with phonology, grammar and semantics, and their relation to problems of language teaching. First of all, however, it may be useful to look more generally at language in comparison with other aspects of behaviour, and then to examine some concepts which are fundamental to the scientific study of language.

2. Some characteristics of human language

I have defined linguistics as the scientific study of language. It would be more accurate to say that linguists are concerned with the study of human language. This does not mean that other living beings do not have forms of language in which they can communicate with members of their own species. Bees and dolphins, for example, are two species which are known to have quite elaborate forms of communication. However, none of the systems of animal communication which have been studied so far possess anything like the flexibility and complexity of human language. What is it, then, that makes human language unique? In an attempt to answer this question, the American linguist C. F. Hockett has proposed a set of "design features" which all human language display. Some of these features are found in the communication systems of other animals, and also in certain human systems(e. g. ,instrumental music and gesture)which do not constitute language in the usual sense of the word. Human language, however, is the only form of communication in which all the design features are combined.

We would realize that among human beings the typical way of transmitting messages is via the speaker's mouth and the listener's ear; in other words, language utilizes the vocal-auditory channel, unlike gestures, the dancing of bees or the courtship ritual of the stickleback. We would also notice that language differs from the ritual of the stickleback with respect to the feature of interchangeability: a speaker of a language can reproduce any linguistic message he can understand, whereas the characteristic courtship motions of the male and female stickleback are different, and neither can act out the motions appropriate to the other.

Continuing our observations, we might conclude that a linguistic message brings about a particular result because there are relatively fixed associations between words and recurrent features of situations in the world around us. For example, when we say "Pass the salt" we generally get what we want because the English word "salt" means salt, not sugar or pepper. Hockett gives the name *semanticity* to this feature of language. Furthermore, we would realize that in language the ties between meaningful elements and their meanings are *arbitrary* and a matter of convention. There is nothing inherently "dog-like" about the word dog; the same animal is called *Hund* in German, *chien* in French, *perro* in Spanish, and *go* in China. Apart from a few "imitative" words which occur in every language (neigh, bleat, crash, etc.) there is no connection between the physical form of a word and what it signifies. A child learning his first language or an adult a foreign language might feel that the feature of arbitrariness of words and language. Since most words in the vocabulary are not "natural," or onomatopoeic, in origin, an effort has to be made to learn each one. But because of the arbitrary nature of the link between message elements and their meanings there is no limit to the number of words in a language, and therefore no limit to what can be talked about.

One of the most important features of language is *productivity*; the capacity of man to say things that have never been said or heard before, and yet to be understood by other speakers. As far as we know only man has the means to coin new utterances by selecting familiar words and phrases, and assembling 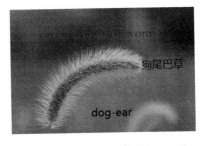 them into new combinations according to rules which are known to all the speakers of a language. In this respect the "open" system of human language differs fundamentally from the gibbon call system, which is "closed", in that each vocal sound is one of a small finite repertoire of fixed, unitary calls.

Displacement, another of Hockett's terms, refers to the ability of man to talk about things that are remote in time and space. This feature occurs in bee dancing but seems to be lacking in the vocal signalling of man's closest living relatives, the gibbons and great apes. It is the faculty of displaced speech that enables man to recount events that happened in the past, to talk about future plans and to create the imaginative events of myth and diction. Displaced speech too has made possible the development of science, since it enables man to accumulate records of his experiences, and to work out problems at leisure without being distracted by the needs of a particular set of circumstances.

In our role as observers of human society, we could not help noticing the importance of *cultural transmission* in the life of a community. Human genes transmit the capacity to acquire language, but the detailed conventions of a particular language—the vocabulary and grammatical rules—are transmitted by teaching and learning. Children are not born with the ability to speak, say, French or English; they have to acquire this ability in the first few years of life, by hearing and imitating what adults say. It is not known to what extent teaching and learning, rather than genetic inheritance, plays a part in gibbon calls or other mammalian systems of vocal signals.

Another important feature of language is *discreteness*. There is virtually no limit to the variety of sound which can be produced by the human vocal organs, but in any one language only a relatively small number of sound-ranges are used, and the difference between these ranges is absolute. For example, the sounds t and d in the English words *seating* and *seeding* are distinguished from one another only with respect to one phonological feature, that of voicing or voicelessness. Both sounds are alveolar stops, but *t* is a member of the class of voiceless stops, while *d* is a member of the class of voiced stops. No speaker will pronounce *seating* and *seeding* exactly alike on all occasions, and it is usual to find considerable variation between speakers in the way these words are pronounced. It is quite likely, for example, that a speaker might produce syllables that deviated from the normal pronunciation of *seating* in the directionof that of *seeding*, or vice versa, and that on some occasions his pronunciation of *seating* is indistinguishable from that of *seeding*. However, in every case he will be understood as saying either *seating* or *seeding*; the principle of discreteness operating in the sound system of English rules out the possibility of a third word containing an alveolar stop with a pronunciation mid-way between seating and seeding.

Linguists studying human speech have established a number of important principles for the description of language structure. For example, it is generally acknowledged that any utterance in any language can be represented as a sequence of distinctive sounds, or phonemes. Thus, the word *give* consists of a sequence of three phonemes /g/ /I/ /v/, *him* consists of three phonemes /h/ /i/ /m/, *a* consists of one phoneme /ə/, and *pen* consists of three phonemes /p/ /e/ /n/. Phonemes have no descriptive meaning in themselves; they serve only to keep meaningful utterances apart, as when *Lend him* a pen is distinguished from Lend him a pin by the difference between the second phonemes of /pen/ and /pin/. Compare also *Send him a pen* with *Lend him a pen*, and *Lend Jim a pin* with *Lend him a pin*. Language also has a structure in terms of minimum meaningful elements, or morphemes. The utterance *He works at night* can be analysed as containing the morphemes he, *work*, -s, at, *night*. It is distinguished from *He worked at night* by the fact that the morpheme - s marking the present tense appears in the first utterance, whereas the morpheme -ed marking the past tense appease in the second utterance.

This feature of language, whereby morphemes are represented by varying arrangements of contrasting sounds which are in themselves meaningless, is called *duality of patterning*. It is duality of patterning that makes it possible for a language to possess thousands of morphemes and to represent them economically by different permutations of a relatively small stock of phonemes, of which there are rarely more than fifty in any one language. The flexibility of language structure is illustrated by the English words *tack*, *cat* and *act*—phonemically /tak/ /kat/ /akt/—which are quite distinct in meaning and composed of three phonemes combined in different ways. It is possible that none of the animal communication systems share this feature of language, certainly none among the other hominids. In a typical animal call system each call differs as a whole from the rest, both in total sound effect and in meaning. There must therefore be a practical limit to the number of distinct messages than can be discriminated, especially when transmission takes place under less than perfect conditions. As a result of duality of patterning, there is no such limit where human language is concerned.

3. Form and meaning

It is a basic principle of linguistics that we should make a clear distinction between a formal analysis of language, and one which is based on meaning. A formal analysis is concerned with the observable, actually occurring forms of language and the relationships between them, while a meaning-based analysis is concerned with the ways in which the forms are used as a vehicle for communication. The views of linguists have differed with respect to the necessity or otherwise of a formal basis for language study. Thus, if we take for granted universal concepts such as 'subject,' 'predicate,' 'dative,' 'locative.' we might feel justified in categorizing the forms of different languages in terms of these concepts, without having to define their meaning afresh each time we embarked on the analysis of a language. On the other hand, we might feel that it is more appropriate to an objective scientific study if we approach each language with a minimum of preconceptions and attempt to explain its regularities only on the basis of observable evidence. The two approaches result in two different types of descriptive framework, neither of which we can afford to neglect, since each has contributed significantly to our understanding of language.

Linguists who adopt a strictly formal approach to the analysis of language aim to establish a set of units which are describable "in physical terms of form, correlations of these forms, and arrangements of order" (Fries 1952). This type of analysis is based on the fact that every linguistic unit below the level of sentence has a characteristic distribution; that is, it is restricted to a greater or lesser degree with respect to the environments in which it can occur. Two or more units occurring in the same range of contexts are said to be distributionally equivalent; if they never occur in the same context they are in complementary distribution. There are certain intermediate cases where the distribution of one unit may include the distribution of another without being totally equivalent to it, or where the distribution of two units overlap but without either of the two occurring in all the contexts where the other occurs, but these need not concern us here. The important point is that in the majority of cases the distribution of units is sufficiently clear-cut to enable the principal grammatical categories to be established without difficulty. It should be noted that distributional equivalence implies grammatical identity only insofar as the contexts are specified by the grammatical regularities of the language. For example, *ate* and *bought* are regarded as distributionally equivalent because they both occur in the context *John _ a cake*, and on these grounds we are able to identify them as members of the same class of lexical items (those usually called "verbs"). Since co-occurrence is established between classes, not items, the classification remains valid even though there are some contexts where *bought* occurs and *ate* does not. For example we have *John bought a car* but not *john ate a car*. The classification of *ate* and *bought* is not affected since the non-occurrence of the second sentence can only be explained in terms of the meaning of the particular words in the sentence, which many structural linguists would regard as being outside the scope of a systematic grammatical statement.

A language has a highly complex structure, and it is impossible for the linguist to describe it all at once. The usual procedure is to divide up the subject-matter into a number of different but interrelated aspects, and to attend to these one at a time. By this means linguists have come to recognize various levels of analysis in the study of language, three of the most commonly discussed levels being those of phonology, syntax and semantics. In addition to various levels the analysis of which may involve different kinds of criteria, we distinguish different ranks at any one level where the same criteria are used to establish units of greater or lesser extent. The sentence is traditionally regarded as the longest structural unit of which a full grammatical analysis is possible, since it is only within a sentence that the interrelations of the elements are completely describable in terms of grammatical rules. It is convenient, therefore, to assume that the domain of grammar is circumscribed by the upper limit of the sentence and the lower limit of the morpheme or minimal grammatical unit. Between these two limits units are abstracted at various ranks and given such names as clause, phrase and word. This arrangement of contrasting categories at successive ranks, the categories at any one rank being included in a category at the next higher rank, is known as a taxonomic hierarchy—a method of classification used in many of the natural sciences. Thus, we may say that sentences consist of one or more clauses, clauses consist of one or more phrases, phrases consist of one or more words, and words consist of one or more morphemes.

Linguistic elements enter into two main types of relation with one another, syntagmatic and paradigmatic. A linguistic element enters into syntagmatic relations with other elements at the same rank with which it forms a serial structure related to linear stretches of writing or the temporal flow of speech. At the same time it enters into paradigmatic relations with other elements which may appear in a given context and which are mutually exclusive in that context. Syntagmatic relations are relations of co-occurrence; paradigmatic relations are relations of substitutability. The first are "overt" relations, realized in the word order of sentences. By contrast, paradigmatic relations are not revealed directly by the observation of any particular sentence, but by comparing a number of similar sentences and ascertaining which elements substitute for one another in a given grammatical context. An illustration should serve to make the distinction clear. In the following diagram the elements in each column are in a paradigmatic relationship to one another. An example of a syntagmatic relationship is that which holds between each of the elements John, met and the vicar in the sentence *John met the vicar*, or which links the vowel-sound /e/ with the /m/ that precedes it and the /t/ that follows it in /met/:

Figure 1

1	2	3
John	met	the vicar
My brother	invited	Mary
He	liked	the Wilsond
Everyone	criticized	our new secretary

The linguist's identification of levels, ranks and form classes is widely exploited in language teaching material. Perhaps the most common type of pedagogic presentation is one in which each "pattern" or grammatical sequence is considered as a row of vacant pigeon-holes into which the grammatical units are sorted in their correct order. Suitable labels may then be given to the contents of each pigeon-hole, e. g. "determiner," "numeral," "adjective," "noun attribute," "noun head." In many textbooks thearrangement of data is governed by a desire to facilitate a process of substitution. According to H. E. Palmer, an "ideal" substitution table is one in which all the elements are mutually intercombinable, as in Figure 1. In other types of tables students must use their own semantic judgement in deciding which elements may be combined. A comparatively simple substitution table can yield a very large number of sentences, all of which belong to a single sentence-type. For example, it is not difficult to construct a table with five columns and five items in each column which will yield thousands of perfectly acceptable sentences (Palmer 1916). It is often said that substitution tables encourage an over-mechanical activity with insufficient scope for the creative intelligence of the student. If there is a fault, however, it would seem to lie not in the concept of substitution tables but in the way such tables are commonly used. There is nothing facile or "meaningless" in the procedure recommended by Palmer, whereby each sentence produced by the substitution table should be "examined, recited, translated, retranslated, acted, thought and concretized".

Turning again to Figure 1 we see that the elements in column 1 are not interchangeable with those in column 2. Nor are the elements in column 2 interchangeable with those in column 3. However, the elements in columns 1 and 3 are interchangeable. Thus we do not normally find *Met John the vicar* or *My brother Mary invited* but we do find *The vicar met John*, *Mary invited everyone*, etc. The fact that nothing in the first column can be substituted for anything in the second column and vice versa shows that the grammatical functions of the elements in the first two columns are quite distinct, and this enables us to set up the two form classes which are traditionally labelled "nouns" abd "verbs", but which Fries, in the interests of objectivity, prefers to call Class I words and Class II words respectively. The fact that all the elements in the third column can substitute in the first column and vice versa enables us to identify all the elements in columns 1 and 3 as members of the same form class. Consequently all the sentences in Figure 1 are examples of a sentence-type in which form classes are arranged in the order Class I+Class II +Class I. This pattern is often referred to as a Subject+Verb+Object, or S+V+O sentence type.

We know that *John loves Mary* does not mean the same thing as *Mary loves John*. This suggests that the elements in column 1 have a different function from the elements in column 3, even though both columns are made up of members of the same form class. The nouns in column 1 function as subjects, while those in column 3 function as objects and have a different relationship with the verb. The notions "subject of", "predicate of", "object of", etc. differ from class names such as "noun" or "verb" in that the former are essentially relational (e. g. , subject and object are differentiated according to the way in which they relate to the verb, while "predicate" is traditionally defined as that part of the sentence which "says something" about the subject). Relational terms such as "subject" and "object" must be handled with care since these expressions can mean different things to different people. For example, in traditional logic the term "subject" refers to a component of a proposition distinguished according to the part it plays in certain

89

rules of inference. Many linguists, on the other hand, define "subject" as the noun, or equivalent word or word group, found in the minimal sentence-type represented by *John works* ("predicate" may then be used to describe the rest of the sentence after the subject has been defined). Others prefer to say that the subject is that part of a sentence which refers to an actor, one who does something, and the object is the part referring to the goal, that which undergoes the action. In many teaching grammars a number of different criteria are used for establishing the identity of subject, predicate and object in different types of sentences. A flexible approach of this type often gives quite good results in the classroom but to avoid confusion it is important that teachers should realize the difference between the various types of definition, and know exactly what criteria are being used at any particular time.

During the past twenty-five years it is the formal aspect of the form-meaning dichotomy that has tended to be dominant in second language teaching. Language teachers have seen it as their main task to give their students a knowledge of the formal, structural patterns of the language being taught. As a result, a typical modern textbook contains plenty of practice in the composition of sentences, but gives little systematic attention to the ways in which the sentences are used for purposes of communication. The assumption is that once the grammatical system has been learned the student will know how to put the system to use in producing sentences of his own, without the need for any further instruction. Even when attempts are made to make the lesson material more meaningful to students, as when the introduction of new structures is associated with explanatory actions and pictures, the highest priority is still given to grammatical criteria, and the artificially-created "situations" bear little resemblance to natural language use.

The so-called audiolingual method of language teaching depends heavily on the use of intensive oral drills, or "pattern practice", designed to give maximum opportunity for practising the structures being taught. Such drills are associated with a high degree of control by the teacher. In a typical syllabus the grammatical patterns are presented one by one, and care is taken to allow the learner plenty of time to absorb each new pattern before he goes on to the next. The aim of this type of teaching is the establishment of automatic speech habits. The exercises are repeated until the student can produce the correct grammatical forms promptly, accurately and with minimum conscious thought. There is no doubt that some aspects of language, especially those involving automatic patterns of concord or "agreement", can be taught very effectively by means of intensive oral drills. Critics of the audiolingual method, however, have pointed out that students tend to become bored by the incessant repetition of formal patterns, especially if no meaningful purpose is apparent in the exercises. Many drills are designed in such a way that the student is able to produce strings of sounds quite mechanically without a thought for the meaning of what is being said. Whatever the intention of such drills may be, their effect is to encourage students to practise the forms of the language, and to neglect the meanings which ought to be associated with the forms.

As an alternative to the pattern-based, habit-formation method, a number of writers have recently proposed a notional or semantic approach to language teaching in which the traditional structural syllabus would be replaced by one based on meaning rather than form. Instead of bringing together sets of grammatically identical sentences, the writer of a notional syllabus would attempt to teach language appropriate to the kind of situation in which the learner is likely to want to use the foreign language. Learning units would have functional rather than grammatical labels; the resulting materials would be functionally unified but grammatically heterogeneous, reflecting how things are in real life where situations do not contain grammatically uniform language. The proposal that text-book writers should pay attention to context in which language is used is of course not new. Generations of tourists visiting foreign countries have equipped themselves with handbooks in which "useful phrases" are collected under headings referring to physical situations— "at the bank", "at the station", "at the theatre"—or to types of functional communication—"ordering a meal", "buying a suit", "asking the way". In order to develop a genuine creative use of language, however, the learner must not confine himself to learning forms solely for their value in a single situation. He needs to study not complete situations but the component parts of situation, not complete sentences but the underlying speech acts—denial, disagreement, affirmation, approval—by means of which we give expression to our views, and attempt to influence the behaviour of others. The aim of the learner should be, not to learn a series of "model conversations" off by heart, but to acquire a set of variable strategies which he can employ for himself as the need arises.

Notes

1. "Some Basic Concepts in Linguistics" consists of ten parts. What you find here are the first three parts. The rest are respectively entitled "Speech and writing," "Language as system," "Decriptive or prescriptive," "Synchronic and diachron studies," "Langue and parole," "Chomsky's theory of language," and "Further readings."

2. These are the chapters in *Papers in Applied Linguistics*.

3. C. F. Hockett: an American linguist who puts forward a set of "design features" for human languages. Hockett is noted for many famous works, among which are *A Course in Modern Linguistics*, *A Manual of Phonology*, *The State of Art*, and *Language, Mathematics and Linguistics*.

Words and Expressions

expound	[ikˈspaʊnd]	vt.	解释,详细讲解;阐述;阐明
dialectical	[ˌdaɪəˈlektɪkl]	adj.	方言的,辩证的;辩证法的
complementation	[ˌkɒmplɪmenˈteɪʃn]	n.	互补,互补作用
canon	[ˈkænən]	n.	标准,准则;教规,宗规
exhaustiveness	[ɪgˈzɔːstɪvnəs]	n.	穷尽性;用尽一切
aesthetically	[esˈθetɪkli]	adv.	审美地,美学观点上地,美学地
lexicography	[ˌleksɪˈkɒgrəfi]	n.	词典编纂
objectivity	[ˌɒbdʒekˈtɪvəti]	n.	客观性,客观现实;客观性原则;客体性
stickleback	[ˈstɪkəlbæk]	n.	棘鱼
semanticity	[səˈmæntəsɪst]	n.	语义学者,语义学家
convention	[kənˈvenʃn]	n.	会议;国际公约;惯例,习俗
dog-like	[ˈdɒglaɪk]	adj.	顽强的,忠实的
onomatopoeic	[ˌɒnəˌmætəˈpiːɪk]	adj.	拟声的,声喻的,拟声词的
gibbon	[ˈgɪbən]	n.	长臂猿
repertoire	[ˈrepətwɑː(r)]	n.	全部节目;全部本领;(计算机的)指令表
unitary	[ˈjuːnətəri]	adj.	单一的;单位的;整体的

displacement	[dis'pleismənt]	n.	取代,替代;免职,置换
ape	[eip]	n.	人猿;猿人;大猩猩;猿
recount	[ri'kaʊnt]	vt.	详细叙述,重新计算,诉说
diction		n.	发音法;措辞,用词;文辞
transmit	[træns'mit]	vt.	传送,传递
inheritance	[in'heritəns]	n.	继承;遗传;遗产 继承性
mammalian	[mæ'meljən]	n.	哺乳动物,哺乳类
		adj.	哺乳动物的,哺乳类的
discreteness	[di'skri:tnəs]	n.	组件,离散性,不连续性
alveolar	[æl'vi:ələ(r)]	n.	齿槽音
phoneme	['fəʊni:m]	n.	音位,音素
morpheme	['mɔ:fi:m]	n.	词素
duality	[dju:'æləti]	n.	二元性,二重性
permutation	[,pə:mju'teiʃn]	n.	序列,排列
hominid	['hɒminid]	n.	类人动物
		adj.	人科的
otherwise	[ʌðəwaiz]	adv/ conj/ conj.	否则,另外;别的,另外的,不然
predicate	['predikət]	vt.	断言,断定;vi. 断言,断定 n. 谓语
dative	['deitiv]	n. & adj.	与格(的)
locative	['lɒkətiv]	n.	(名词、代词和形容词的)方位格
afresh	[ə'freʃ]	adv.	重新
embarked on	[im'bɑ:k ɒn]		着手,开始做某事
preconception	[,pri:kən'sepʃn]	n.	先入之见,偏见
complementary	[,kɒmpli'mentri]	adj.	互补的;补充的
intermediate	[,intə'mi:diət]	adj.	中间的 clear-cut
insofar	[,insəʊ'fɑ:]	adv.	在这个范围,到这种程度
co-occurrence			共现;共现关系
taxonomic	[,tæksə'nɒmik]	adj.	分类学的
hierarchy	['heɑ:ki]	n.	分层,层次;等级制度;阶层;层次结构
syntagmatic	[,sintæg'mætik]	adj.	组合关系的;横组合的
paradigmatic	[,pærədig'mætik]	adj.	词形变化(表)的,范例的,聚合关系的
substitutability	[,səbstə,tjutə'biləti]	adj.	可代换性,可置换性

overt [əʊˈvɜːt]	adj.	明显的;公然的;公开的
ascertain [ˌæsəˈtein]	vt.	弄清,确定,查明
vicar [ˈvikə(r)]	n.	(英国国教的)教区牧师;教皇
pedagogic [ˌpedəˈgɒdʒik]	adj.	教师的,教育学的,教授法的
facile [ˈfæsail]	adj.	轻率作出的;不动脑筋的,肤浅的
concretize [ˈkɒnkriːtaiz]	vt.	使具体化,使有形化
proposition [ˌprɒpəˈziʃn]	n.	命题;建议;主张
dichotomy [daiˈkɒtəmi]	n.	一分成二,对分;二分法;两分法
resemblance [riˈzembləns]	n.	相似,形似;相似物
audiolingual [ˌɔːdiəʊˈliŋgwəl]	adj.	(语言教学)听说的,听说法

Questions for Discussion and Review

1. According to J. P. B. Allen, what is linguistics? In what sense is the study of language scientific?

2. Does the writer agree with Chomsky that "This creative aspect of normal language use is one fundamental factor that distinguishes human language from any known system of animal communication"? Do you agree? Defend your answer.

3. Explain the functions of the various kinds of rules known by a speaker of English (or of any language).

4. Explain the concepts of "closeness" and "remoteness." Why are they important?

5. How does the writer define the "syntagmatic" and "paradigmatic" relations? What are their functions? Give some examples of your own to explain how they are used in the arrangement of language elements.

6. Explain as fully as you can how J. P. B. Allen demonstrates the importance of cultural transmission in language teaching and learning.

Unit 9

Phonology

> The research interests of George N. Clements relate to the representation of the phonetic, phonological and prosodic aspects of linguistic knowledge. He writes books and articles about the various kinds of relationship that exist between the sounds we use. Human languages display a great many speech sounds, and phonology, a component of grammar, determines how sound patterns are formed in a language. In the following article, Clements first defines phonology, phoneme, phonological features, rules and processes, and then sketches out the scope of development in phonological research.

Phonology addresses the question of how the words, phrases, and sentences of a language are transmitted from speaker to hearer through the medium of speech. It is easy to observe that languages differ considerably from one another in their choice of speech sounds and in the rhythmic and melodic patterns that bind them together into units of structure and sense. Less evident to casual observation, but equally important, is the fact that languages differ greatly in the way their basic sounds can be combined to form sound patterns. The phonological system of a given language is the part of its grammar that determines what its basic phonic units are and how they are put together to create intelligible and natural-sounding spoken utterances.

Let us consider what goes into making up the sound system of a language. One ingredient, obviously enough, is its choice of speech sounds. All languages deploy a small set of consonants and vowels, called phonemes, as the basic sequential units from which the minimal units of word structure are constructed. The phonemes of a language typically average around 30, although many have considerably more or less. English has about 43, depending on how we count and what variety of English we are describing. Although this number may seem relatively small, it is sufficient to distinguish the 50,000 or so items that make up the normal adult LEXICON. This is due to the distinctive role of order: thus, for example, the word step is linked to a sequence of phonemes that we can represent as /step/, whereas pest is composed of the same phonemes in a different order, /pest/.

Phonemes are not freely combinable as a maximally efficient system would require but are sequenced according to strict patterns that are largely specific to each language. One important organizing principle is *syllabification*. In most languages, all words can be exhaustively analyzable into syllables. Furthermore, many languages require all their syllables to have vowels. The reason why a fictional patronymic like Btfsplk is hard for most English speakers to pronounce is that it violates these principles—it has no vowels, and so cannot be syllabified. In contrast, in one variety of the Berber language spoken in Morocco, syllables need not have vowels, and utterances like tsgssft stt ("you shrank it") are quite unexceptional. Here is a typical, if extreme, example of how sound patterns can differ among languages.

Speech sounds themselves are made up of smaller components called DISTINCTIVE FEATURES, which recur in one sound after another. For example, a feature of tongue-front ARTICULATION (or coronality, to use the technical term) characterizes the initial phoneme in words like *tie, do, see, zoo, though, lie, new, shoe, chow,* and *jay*, all of which are made by raising the tip or front of the tongue. This feature minimally distinguishes the initial phoneme of *tie* from that of *pie*, which has the feature of labiality (lip articulation). Features play an important role in defining the permissible sound sequences of a language. In English, for instance, only coronal sounds like those just mentioned may occur after the diphthong spelled *ou* or *ow*: We consequently find woeds like *out, loud, house, owl, gown,* and *ouch,* all words ending in coronal sounds, but no words ending in sound sequences like *owb, owf, owp, owk,* or *owg*. All speech sounds and their regular patterns can be described in terms of a small set of such features.

A further essential component of a sound system is its choice of "suprasegmental" such as LINGUISTIC STRESS, by which certain syllables are highlighted with extra force or prominence; TONE, by which vowels or syllables bear contrastive pitches; and intonation, the overall "tune" aligned with phrases and sentences. Stress and tone may be used to distinguish different words. In some varieties of Cantonese, for example, only tone distinguishes si "poem" (with high pitch), si "cause" (with rising pitch), and si "silk" (with falling pitch). Prosodic features may also play an important role in distinguishing different sentence types, as in conversational French where only intonation distinguishes the statement *tu viens* "you come" (with falling intonation) from the corresponding question *tu viens?* (with rising intonation). In many languages, stress is used to highlight the part of the sentence that answers a question or provides new information (cf. FOCUS). Thus in English, an appropriate reply to the question "Where did Calvin go?" is *He went to the STORE*, with main stress on the part of the sentence providing the answer, whereas an appropriate reply to the question "Did you see Calvin with Laura?" might be No, *I saw FRED with* her, where the new information is emphasized. Though this use of stress seems natural enough to the English speaker, it is by no means universal, and Korean and Yoruba, to take two examples, make the same distinctions with differences in word order.

Although phonological systems make speech communication possible, there is often no straightforward correspondence between underlying phoneme sequences and their phonetic realization. This is due to the cumulative effects of sound changes on a language, many of them ongoing, that show up not only in systematic gaps such as the restriction on vowel+consonant sequences in English noted above but also in regular alternations between different forms of the same word or morpheme. For example, many English speakers commonly pronounce *fields* the same way as *feels*, while keeping *field* distinct from *feel*. This is not a matter of sloppy pronunciation but of a regular principle of English phonology that disallows the sound [d] between [l] and [z]. Many speakers of American English pronounce *sense* in the same way as *cents*, following another principle requiring the sound [t] to appear between [n] and [s]. These principles are fully productive in the sense that they apply to any word that contains the underlying phonological sequence in question. Hosts of PHONOLOGICAL RULES AND

PROCESSES such as these, some easily detected by the untrained ear and others much more subtle, make up the phonological component of English grammar, and taken together may create a significant "mismatch" between mentally represented phoneme sequences and their actual pronunciation. As a result, the speech signal often provides an imperfect or misleading cue to the lexical identity of spoken words. One of the major goals of speech analysis—one that has driven much research over the past few decades—is to work out the complex patterns of interacting rules and constraints that define the full set of mappings between the underlying phonemic forms of a language and the way these forms are realized in actual speech.

Why should phonological systems include principles that are so obviously dysfunctional from the point of view of the hearer (not to mention the language learner)? The answer appears to lie in the constraints imposed "from above" by the brain and "from below" by the size, shape, and muscular structure of the speech-producing apparatus (the lungs, the larynx, the lips, and the tongue). The fact that languages so commonly group their phonemes into syllables and their syllables into higher-level prosodic groupings (metrical feet, phrases, etc.) may reflect a higher-order disposition to group serially ordered units into hierarchically organized structures, reflected in many other complex activities such as memorization, versification (see METER AND POETRY), and jazz improvisation. On the other hand, human biology imposes quite different demands, often requiring that complex phonemes and phoneme sequences be simplified to forms that are more readily articulated or that can be more easily distinguished by the ear.

Research on phonology includes the development of so-called nonlinear (auto-segmental, metrical, prosodic) models for the representation of tone, stress, syllables, feature structure, and prosodic organization, and the study of the interfaces between phonology and other areas of language, including SYNTAX, MORPHOLOGY, AND PHONETICS. At the present time, newer phonological models emphasizing the role of constraints over rewrite rules have become especially prominent, and include principles-and-parameters models, constraint-and-repair phonology, declarative phonology, connectionist-inspired approaches, and most recently OPTIMALITY THEORY.

Viewed from a cognitive perspective, the task of phonology is to find the mental representations that underlie the production an perception of speech and the principles that relate these representations to the physical events of speech. This task is addressed hand-in-hand with research in related areas such as *LANGUAGE ACQUISITION* and language pathology, acoustic and articulatory phonetics, *PSYCHOLINGUISTICS*, neurology, and computational modeling. The next decades are likely to witness increased cross-collaboration in these areas.

As one of the basic areas of grammar, phonology lies at the heart of all linguistic description. Practical applications of phonology include the development of orthographies for unwritten languages, literacy projects, foreign language teaching, speech therapy, and man-machine communication.

Words and Expressions

prosodic [prəʊˈsɒdik]	adj.	韵律的；节律的
exhaustively [igˈzɔːstivli]	adv.	用尽一切地
fictional patronymic [ˌpætrəˈnimik]		虚构的父名
shrank [ʃræŋk]	v.	收缩(shrink 过去式)(使)缩水；退缩；畏缩
labiality [leibiˈæliti]	n.	唇音性
coronal [kəˈrəunəl]	n.	花冠，冠状物，前舌音
aligned with [əˈlain wið]	v.	与…结盟
cantonese [ˌkæntəˈniːz]	n.	广东人，广东话
yoruba [ˈjɔːrəbə]	n.	(非洲)约鲁巴人(语)
sloppy [ˈslɒpi]	adj.	草率的；懒散的，拖泥带水
dysfunctional [disˈfʌŋkʃənl]	adj.	功能失调的
larynx [ˈlæriŋks]	n.	喉
serially [ˈsiəriəli]	adv.	连续地，一系列地
hierarchically [ˌhaiəˈrækikəli]	adv.	分等级地，分级体系地
versification [ˌvəːsifiˈkeiʃn]	n.	诗律，作诗
improvisation [ˌimprəvaiˈzeiʃn]	n.	即兴创作，即兴演说
interface [ˈintəfeis]	n.	界面；<计>接口；交界面
parameter [pəˈræmitə(r)]	n.	参数，参量；限制因素；决定因素
declarative phonology		陈述式音系学
connectionist [kəˈnekʃənist]		连接主义
OPTIMALITY [ˌɒptiˈmoeliti]	n.	最优性；最佳化

perspective	[pəˈspektiv]	n.	景色;洞察力,视角
pathology	[pəˈθɒlədʒi]	n.	病理(学);〈比喻〉异常状态
neurology	[njʊəˈrɒlədʒi]	n.	＜神经病学;神经病学的,神经学
collaboration	[kəˌlæbəˈreiʃn]	n.	合作,协作
orthographie	[ɔːˈθɒɡrəfi]	n.	正确拼字,正字法,正字学,表音法
therapy	[ˈθerəpi]	n.	治疗,疗法,疗效
matrix	[ˈmætrisiːz]	n.	母体,子宫;矩阵,本体

Questions for Discussion and Review

1. Why does phonology focus on speech and not on writing? Why is a phonetic alphabet necessary for work in phonology?

2. Define phonemes and syllabification.

3. Give three English words where stress distinguishes between verb and noun.

4. Explain why somebody might choose to stress the following utterances as indicated by the bold type:

John wanted to do this today.

John **wanted** to do this today.

John wanted to do this **today**.

5. Only six consonants can precede [1] word initially:

[p] as in play　　　[k] as in clean　　　[s] as in slip

[b] as in black　　　[g] as in glow　　　[f] as in flip

Write feature matrices which describe the classes of sounds that can occur before [I].

Unit 10

Phonetic Distinctive Features

> Robert E. Callary, a professor of Northern Illinois University, is concerned with linguistic theories and dialectology, the branch of linguistics that deals with regional and social variation in language. It is a basic principle of linguistics that language is rule-governed. The phonological component must include the rules that convert syntactic units into sound units. A description of the sounds of a language, of the rules governing the patterning of those sounds, and of the functioning of the various phonological processes of the language constitutes the phonology or phonological component of the grammar of the language. In the following article, Callary explains how the sounds of English are analyzed and transcribed, and discusses phonetic distinctive feature analysis. On this basis, he further examines some common rules of English and some key phonological processes.

Languages consist of several interrelated systems called components. These include a semantic component which determines the meanings of words and sentences, a syntactic component which specifies how sentences may be created or changed, and a Phonological component which contains the rules by which syntactic units are converted into sound units. Since sound is the primary vehicle for expressing language, the phonological system is of major importance in understanding the nature of language and the nature of our knowledge of language.

Syntax looks like a tree? A nature classification?

Phonology, the sound system of language, is usually divided into two parts: the elements of the sound system, and the organizational patterns of these elements. The science of phonetics has as its main objective: the identification and description of the sounds found in language.

Phonology is also concerned with the "grammar" of sounds: the patterns they enter into and the changes they undergo when juxtaposed with other sounds in the course of normal speech. For example, the grammar of English phonology allows a maximum of three consonants to cluster at the beginning of a word - but only a very few consonants are permitted in this position, and then only in a certain order. The first must be [s], the second must be one of [p], [t], or [k], and the third either [I] or [r] (splint, string, spring, screen, etc.). All other arrangements, such as *trs-* and *rst-*, are prohibited by the rules of English phonology.

Phonological rules also specify how sounds regularly change when spoken in a variety of contexts. Notice how the final [t] of words *moist* changes when the suffix *-ure* is added (*moisture*), and how the final [k] of *public* changes to [s] before the suffix *-ity* (*publicity*).

The object of phonology, and of other areas of language study as well, is to describe and ultimately to explain the knowledge speakers have which allows them to produce and to understand their language. This knowledge is called linguistic competence. In phonology, competence includes knowledge of the specific sounds that occur in a language and how these sounds may be strung together to form syllables, words, and longer utterances. Part of competence is knowing what is permissible in a language and what is not. Speakers of English "know," since it is part of their competence, that the initial sequence *str-* is possible, but that other combinations, such as *trl-* and *stm-* are not. By far the greatest part of linguistic competence is unconscious; that is, it is difficult if not impossible to state overtly what the principles of language are. Phonology is an attempt to make explicit one aspect of this linguistic knowledge.

1. Phonetics

The purpose of phonetics is to provide an inventory and a description of the sounds found in speech. There are several ways of looking at these sounds: first, sounds as generated by a source; second, sounds as transmitted through a medium; and third, sounds as perceived by a receiver. In human communication, the first approach describes speech sounds as products of the vocal apparatus which produced them, the second describes the nature of the sound waves resulting from speech, and the third analyzes speech sounds as they strike the listener's eardrum and are interpreted by the brain. The study of speech as sound waves is acoustic phonetics, and the study of the reception of speech sounds is auditory or perceptual phonetics. These two branches of phonetics require sophisticated equipment to analyze and measure the components of speech and, therefore, they are appropriately found in the phonetics laboratory. Articulatory phonetics, however, studies the way or ways a sound is articulate by a speaker, the assumption being that individual qualities of speech sounds result from the particular configurations of the vocal apparatus as the sounds are being produced.

One important point must be made here. When we talk of the sounds of speech, we mean just that—not the way they happen to be represented on a printed page. To confuse sound and spelling is to confuse language with its representation. In many languages, the relationship between sound and spelling is very close. In these languages we have a very good chance of guessing the correct pronunciation of a word on the first try. There are many words in English, however, which are spelled in such a way as to conceal their correct pronunciation. This ill fit in many words between their sound and their spelling is the result of three factors: there may be more (or fewer) sounds in a word than the spelling would lead us to believe, the same sound may be represented by a variety of spellings, and the same spelling may represent several different sounds. For example:

Same sound /ei/ with different spelling	Same (o) spelling with different sound
Able [A]	cold [kəʊld]
Wait [ai]	cot [kɑːt]
Weight [ei]	corn [kɔːn]

Day [ay]   love [lʌv]
Obey [ey]   wolf [wʊlf]
Gauge [au]   prove [pruːv]
Steak [ea]   moss [mɒs]

Sound, and not spelling, is important in understanding how phonology functions. For instance, there is good reason to group together *cool*, *character*, and *keen*, since they begin with the same sound, even though that sound is spelled different in each word. On the other hand, there is no reason to group together *cool*, *city*, and *cello*, even though they begin with the same letter.

2. Phonemes, Allophones and Allophonic Rules

Phonology has two levels. Phonetics is the physical and concrete level, since it describes sounds exclusively in terms of physical phenomena. The psychological or mental aspect of phonology is the abstract level, where sounds are treated as mental images rather than as physical facts.

It is a curious fact of phonology that speakers consistently recognize and interpret physically different sounds as the "same" sound. Since it is practically impossible to pronounce words, even the same word, in exactly the same way twice, an incredible variety of speech sounds reaches our ears every day. One reason for this variety is that the size and shape of the vocal tract varies from one speaker to the next, and, since speech sounds derive from anatomical functions, different anatomies produce different sounds. Due to vocal tract size alone, the word *me* will be pronounced differently by an adult male, an adult female, and a child. However, these differences do not affect meaning. *Me* is easily recognized as *me*, and not *see*, or *tea*, or something else.

More important from a linguistic point of view than differences due to anatomy are differences which are part of the phonological system and are therefore consistent from one speaker to the next. These variations within sounds are generally determined by the phonetic environment in which the sound occurs. This means that a given sound will have several pronunciations, or variants. Speakers know which variant occurs in which context. This knowledge is part of speakers' phonological competence; it is something they do habitually and instinctively.

As an example of the different phonetic forms a sound may take, consider [t] as it occurs in *top*, *stop*, *pit*, *mutton*, *eighth*, *startle*, and *city*. Functionally, each of these occurrences of [t] is the "same." Physically, each is different. The [t] of top is produced with an accompanying puff of air (called aspiration), while the [t] of stop is unaspirated. (You can easily feel aspiration for yourself by placing your fingertips against your lips and alternately pronouncing *top and stop*.) The final sound of *pit* maybe pronounced in several ways; it may be aspirated or unaspirated, or it may even be unreleased. A sound is unreleased when we make closure between the tongue tip and alveolar ridge, but stop articulating before the sound is released. We have the option of non-release only at the ends of words.

The air stream may exit through either the oral or the nasal cavities. Usually [m, n, and ŋ] are the only sounds released through the nose. All others have an oral, except when they precede a nasal. The [t] of mutton has a nasal rather than an oral release. In articulating mutton, we close for [t] at the alveolar ridge, but before breaking the contact, we drop the velum and release both [t] and the following nasal through the nasal cavity. Other words having a nasally released [t] include *rotten*, *tighten*, *shorten*, *Latin*, and *bitten*.

These four varieties of t are distinguished from each other by manner of articulation differences (aspiration, no aspiration, unrelease, nasal release). The [t] of *eighth*, however, differs from the others in its point of articulation. This [t] is made not at the alveolar ridge as are the [t]'s of *top*, *stop*, *pit*, and *mutton*, but with the tongue tip between the teeth. Since the teeth are involved, this sound has a dental articulation.

The second [t] of startle has two characteristic features. It is voiced rather than voiceless, making it more like a [d] than a [t]. The [t] of *city* is voiced, but its manner of articulation distinguishes it from the [t] of *startle*. In city (and in party, water, and butter), the medial consonant is produced by bouncing the tongue tip off the alveolar ridge. Sounds articulated in this manner are called flaps. Flaps are always voiced, and they occur medially for spelled *t* and *d*, resulting in such pairs as *latter/ladder*, *atom/Adam*, and *traitor/trader* which, for most speakers, are pronounced exactly alike.

Speakers know intuitively when to use each variant of a sound. Since the description of phonology should correspond to speakers' intuitive knowledge, it, too, must include this information. The phonological rules must in effect "tell" us how to pronounce a given sound in a variety of environments. Concerning the examples above, these rules must tell us how to correctly pronounce the different [t]'s of *top*, *stop*, *pit*, *mutton*, *eighth*, *startle*, and *city*. Phonetics would tell us that the [t] of *top* is aspirated, that of *mutton* is nasally released, that of *startle* is voiced, and so forth. But while a phonetic description accounts for our knowledge of the physical makeup of speech sounds, it does not explain why we consider phonetically different sounds to be "the same." We realize that in spite of obvious physical differences, each of the [t]'s above is the "same" sound. Different sounds are the same sound; this seems to be a contradiction.

The contradiction is more apparent than real, however, but to understand it requires us to go beyond the phonetic reality of the composition of speech sounds to the mental realities of language users, to the intentions of speakers, and to the interpretation and classification of speech sounds by hearers.

The phonetic level of phonology describes the minute physical differences we must consistently make in order correctly to pronounce a sound in a given environment; the mental level includes only the information necessary to identify and classify the sound. Speakers intend a certain sound (mental level) and then articulate it (physical level). Listeners hear a sound (physical level) and then interpret and classify it (mental level). If, for example, speakers were to articulate the phrase "top button," they would feel that they were producing the same [t] in each word—not any particular kind of [t], just [t]. But one [t] may very well be different than the next because of the rules of English phonology. The [t] of top must be aspirated and the [t] of button must be nasally released. Speakers, knowing these rules, automatically aspirate the [t] of top and nasally release the [t] of button. On their part, listeners hear different [t]'s in *top* and *button*, but classify them as "the same."

A single mental unit, in this case /t/, may have several physical phonetic units corresponding to it. A mental unit of speech is called a phoneme. The physical units, the actual sounds of speech, are called allophones. In the words *top*, *stop*, *pit*, *mutton*, *eighth*, *startle*, and *city*, there is only one /t/ phoneme, but there are six allophones—aspirated [t], unaspirated [t], unrealsed [t], nasally released [t], laterally released [t], and dental [t]. Phonemes are abstractions; they never occur as such; rather, they are manifested in speech by one or more allophones. Phonemes are units of our intention and our interpretations; they are what we think we utter and what we think we hear. But phonemes are neither spoken nor heard. We actually utter and actually hear one of the allophones of a phoneme rather than the phoneme itself.

A major difference between phonemes and allophones lies in the respective roles they play within the phonological system of the language. Phonemes are contrastive sound units: they make a difference in words. Substituting one phoneme for another will usually result in a word of another meaning. /t/ and /d/ are different phonemes because *time* is a different word than *dime*. Allophones, on the other hand, are non-constrastive. Substituting one allophone for another will not result in a different word, but rather the same word with an unusual pronunciation. If, for example, a person were to pronounce *city* with an aspirated [t] allophone rather than the flap allophone, we would still recognize the word as *city*, but we might wonder whether English was the speaker's native language.

Phonemes can be easily identified because of their contrastive function. Since *time* and *dime* are different words, we know that /t/ and /d/ are phonemes rather than allophones. Words such as *time* and *dime*, which differ only in a single phoneme, are called minimal pairs. Larger groups of words which show phonemic distinctions are called minimal sets. *Time*, *dime*, *rhyme*, *lime*, *chime*, and *mime* constitute a minimal set. From this set we can identify six phonemes: /t/, /d/, /r/, /l/, /č/ and /m/. Since the phoneme is contrastive, it can be defined as a sound unit which constitutes the only difference between minimal pairs.

While phonemes are contrastive, allophones are complementary, meaning that they are distributed in an orderly manner, and we can in most cases predict where an allophone will occur. This is because allophones are usually restricted to a particular environment within a word: one allophone will occur in one position and other allophones will occur only in other positions. The aspirated allophone of /t/ (and of /p/ and /k/ as well) occurs at the beginning of a syllable (tan, pan, can); unaspirated, unreleased, laterally released, nasally released, and flap allophones cannot occur here. Unaspirated allophones occur only after /s/ (stan, span, scan), a position where aspirated allophones cannot occur. Other allophones are similarly restricted. Unreleased [t] occurs only word-finally, laterally released [t] only preceding [l], nasally released [t] only preceding [n], and flap [t] only between vowels.

Though the distribution of allophones is usually complementary, there are times when any of several allophones may occur in the same position within a word. When this happens, the allophones are in "free variation." Word-final is generally a position which favors free variation. Aspirated /t/ unaspirated /t/, or even unreleased /t/ are all possible; they vary freely in this environment.

All phonemes have several allophones, and since the relationships between phonemes and their allophones are generally predictable, they can be described by rules. Rules which convert abstract phonemes into concrete, pronounceable forms are called allophonic rules. We know intuitively that /t/ is aspirated initially, unaspirated after /s/, given a lateral release before /l/ and so forth. This knowledge is made explicit by allophonic rules.

Speech is a continuum of sound rather than a discrete series of sounds one following neatly after another. One sound blends into the next, and influences those sounds on either side of it and it in turn is influenced by the sounds which follow and precede it. Some argue that everyone should strive toward the "correct" pronunciation of words, in which each sound is given its "true" value each time it occurs. But such would be contrary to the nature of language.

3. Phonological Process Rules

Phonological rules describe the regularities of the sound system of language. These rules are of two kinds: allophonic rules, which describe the possible pronunciations of a phoneme, and process rules, which add, delete, or change entire phonemes. Process rules describe the possible pronunciations of a morpheme.

A morpheme is the smallest meaningful part of a word. Some words, such as *apple* and *mayonnaise*, consist of only a single morpheme, while others, such as *pacify and untruths*, consist of several. *Pacify* is composed of a base, *pac-*, a variant of *peace*, and a suffix, *-ify*, which changes the noun peace into the verb *pacify*. Untruths is a three morpheme word: the prefix *un-*, meaning "not," the base truth, and the suffix *-s*, meaning "plural." It is a commonplace of language that a morpheme does not always appear in the same form. On one occasion it will be represented by one string of phonemes and on another by a different string. And in many cases, the various phonemic forms of a morpheme are predictable by phonological rules.

The plural morpheme has three regular phonemic variants, called allomorphs. They are distributed as follows:

/s/ is added to *safe*, *back*, *bath*, etc.

/z/ is added to *save*, *bag*, *tub*, etc.

/(z/ is added to *bus*, *buzz*, *bunch*, *judge*, *bush*, etc.

These allomorphs are phonologically determined; that is, the particular allomorph which occurs is determined by the phonetic makeup of the singular noun to which it is attached. Here voicing assimilation is at work. The voiceless allomorph /s/ is used to form the plural when the final sound of the voiced allomorph /z/ is added to singular nouns ending in voiced sounds. Since /z/ occurs in two of the allomorphs, we will use it to represent the plural morpheme and then write rules to derive the other two allomorphs.

There is much more to phonology than is contained in these few pages; of the kinds of rules needed for a full description of the sound patterns of a language this has been only a sampling. Moreover, as modern phonology is one of the newer branches of linguistics, there is some disagreement among phonologists as to how best to describe phonological phenomena. All phonologists agree, however, that there are two levels to phonology: an abstract, psychological level, and a concrete, physical level. Mediating these levels are phonological rules, which convert a sequence of phonemes into a sequence of pronounceable phones. The exact formulation of these rules differs from one phonologist to the next. But they all have as their goal the description of speakers' knowledge of the sound systems of their language.

Notes

1. In this article, linguistic units are enclosed in brackets in order to distinguish them from spelling.

2. "Eighth" has two regular pronunciations: [etθ] and [eθ]. This discussion is of the [etθ] pronunciation only.

3. In this article, allophones (phonetic units) are enclosed in [ ], phonemes (contrastive units) in / /.

Words and Expressions

juxtapose   [ˌdʒʌkstəˈpəʊz]   vt.   把…并置,并置;并列;并排;使并

			置
be strung together			被串在一起
overtly	[ˈəʊvəːtli]	adv.	明显地,公然地;公开地
explicit	[ikˈsplisit]	adj.	清楚的;详述的;明白的
inventory	[ˈinvəntəːri]	n.	存货清单;编制的目录;清单;总结
Phoneme	[ˈfəʊniːm]	n.	音位,音素
Allophone	[ˈæləfəʊn]	n.	音位变体
anatomical	[ˌænəˈtɒmikl]	adj.	结构(上)的,解剖的,解剖学的.
aspiration	[ˌæspəˈreiʃn]	n.	强烈的愿望;发送气音
alveolar	[ælˈviːələr]	n.	齿槽音
nasal	[ˈneizl]	n.	鼻音字母,鼻音
velum	[ˈviːləm]	n.	膜,软腭
startle	[ˈstɑːtl]	vt./vi.	惊跳,使大吃一惊;使惊跳
		n.	惊愕;惊恐;惊跳
medial	[ˈmiːdiəl]	adj.	中间的,平均的
flap		n.	拍打,拍打声
lime	[laim]	n.	酸橙;石灰;绿黄色;椴树
chime	[tʃaim]	n.	合奏钟声,钟乐,谐音,韵律,和谐
strive	[straiv]	vi.	努力奋斗,力求;斗争
mayonnaise	[ˌmeiəˈneiz]	n.	蛋黄酱
allomorph	[ˈæləmɔːrf]	n.	词素变体;形位变体

Questions for Discussion and Review

1. Describe the subfields of phonetics.

2. What three aspects of articulation must be included in the description of English consonants? What aspects of articulation are included in an articulatory description of English vowels?

3. What are some of the advantages of phonetic distinctive feature analysis as compared to traditional articulatory descriptions? What is a "natural class"? Explain why the concept is useful.

4. Define assimilation and give some examples not described in the text. Why is it so prevalent in speech? Define dissimilation. What function(s) does it serve?

5. Phonological processes:

Clusters of three consonants are not usually allowed in English. Words ending in -st, for example, usually undergo some change whenever a suffix beginning with a consonant is added. "Christmas" becomes [krismas], "post + man" becomes [posman], "waste + paper" becomes [wespepar]. State the process at work in these examples and explain why it is a natural process.

Unit 11

Form and Meaning in Natural Languages

> Noam Chomsky, Professor at the Department of Linguistics and philosophy of Massachusetts Institute of Technology, revolutionized linguistics with his innatist theory of language. Transformational-generative grammar was first brought extensive attention in Chomsky's 1957 book *Syntactic Structures*. This article from *Language and Mind* (1968) explains why he believes the study of human language is so important, and explains and defends some of his assumptions and goals. Viewing the study of human language as "a branch of theoretical human psychology," he argues that "Its goal is to exhibit and clarify the mental capacities that make it possible for a human to learn and use a language. As far as we know, the capacities are unique to man…"

When we study human language, we are approaching what some might call the "human essence," the distinctive qualities of mind that are, so far as we know, unique to man and that are inseparable from any critical phase of human existence, personal or social. Hence the fascination of this study, and, no less, its frustration. The frustration arises from the fact that despite much progress, we remain as incapable as ever before of coming to grips with the core problem of human language, which I take to be this: Having mastered a language, one is able to understand an indefinite number of expressions that are new to one's experience, that bear no simple physical resemblance and are in no simple way analogous to the expressions that constitute one's linguistic experience; and one is able, with greater or less facility, to produce such expressions on an appropriate occasion, despite their novelty and independently of detectable stimulus configurations, and to be understood by others who share this still mysterious ability. The normal language use is, in this sense, a creative activity. This creative aspect of normal language use is one fundamental factor that distinguishes human language from any known system of animal communication.

It is important to bear in mind that the creation of linguistic expressions that are novel but appropriate is the normal mode of language use. If some individual were to restrict himself largely to a definite set of linguistic patterns, to a set of habitual responses to stimulus configurations, or to "analogies" in the sense of modern linguistics, we would regard him as mentally defective, as being less human than animal. He would immediately be set apart from normal humans by his inability to understand normal discourse, or to take part in it in the normal way— the normal way being innovative, free from control by external stimuli, and appropriate to new and ever changing situations.

It is not a novel insight that human speech is distinguished by these qualities, though it is an insight that must be recaptured time and time again. With each advance in our understanding of the mechanisms of language, thought, and behavior, comes a tendency to believe that we have found the key to understanding man's apparently unique qualities of mind. These advances are real, but an honest appraisal will show, I think, that they are far from providing such a key. We do not understand, and for all we know, we may never come to understand what makes it possible for a normal human intelligence to use language as an instrument for the free expression of thought and feeling; or, for that matter, what qualities of mind are involved in the creative acts of intelligence that are characteristic, not unique and exceptional, in a truly human existence.

I think that this is an important fact to stress, not only for linguists and psychologists whose research centers on these issues, but, even more, for those who hope to learn something useful in their own work and thinking from research into language and thought. It is particularly important that the limitations of understanding be clear to those involved in teaching, in universities and schools. There are strong pressures to make use of new educational technology and to design curriculum and teaching methods in the light of the latest scientific advances. In itself, this is not objectionable. It is important, nevertheless, to remain alert to a very real danger: that new knowledge and technique will define the nature of what is taught and how it is taught, rather than contribute to the realization of educational goals that are set on other grounds and in other terms. Let me be concrete. Technique and even technology is available for rapid and efficient inculcation of skilled behavior, in language teaching, teaching of arithmetic and other domains. There is, consequently, a real temptation to reconstruct curriculum in the terms defined by the new technology. And it is not too difficult to invent a rationale, making use of the concepts of "controlling behavior," enhancing skills, and so on. Nor is it difficult to construct objective tests that are sure to demonstrate the effectiveness of such methods in reaching certain goals that are incorporated in these tests. But successes of this sort will not demonstrate that an important educational goal has been achieved. They will not demonstrate that it is important to concentrate on developing skilled behavior in the student. What little we know about human intelligence would at least suggest something quite different: that by diminishing the range and complexity of materials presented to the inquiring mind, by setting behavior in fixed patterns, these methods may harm and distort the normal development of creative abilities. I do not want to dwell on the matter. I am sure that any of you will be able to find examples from your own experience. It is perfectly proper to try to exploit genuine advances in knowledge, and within some given field of study, it is inevitable, and quite proper, that research should be directed by considerations of feasibility as well as considerations of ultimate significance. It is also highly likely, if not inevitable, that considerations of feasibility and significance will lead in divergent paths. For those who wish to apply the achievements of one discipline to the problems of another, it is important to make very clear the exact nature not only of what has been achieved, but equally important, the limitations of what has been achieved.

I mentioned a moment ago that the creative aspect of normal use of language is not a new discovery. It provides one important pillar for Descartes' theory of mind, for his study of the limits of mechanical explanation. The latter, in turn, provides one crucial element in the construction of the anti-authoritarian social and political philosophy of the Enlightenment. And, in fact, there were even some efforts to found a theory of artistic creativity on the creative aspect of normal language use. Schlegel, for example, argues that poetry has a unique position among the arts, a fact illustrated, he claims, by the use of the term "poetical" to refer to the element of creative imagination in any artistic effort,—as distinct, say, from the term "musical," which would be used metaphorically to refer to a sensual element. To explain this asymmetry, he observes that every mode of artistic expression makes use of a certain medium and that the medium of poetry—language—is unique in that language, as an expression of the human mind rather than a product of nature, is boundless in scope and is constructed on the basis of a recursive principle that permits each creation to serve as the basis for a new creative act. Hence the central position among the arts of the art forms whose medium is language.

The belief that language, with its inherent creative aspect, is a unique human possession did not go unchallenged, of course. One expositor of Cartesian philosophy, Antoine Le Grand, refers to the opinion "of some people of the East Indies, who think that Apes and Baboons, which are with them in great numbers, are imbued with understanding, and that they can speak but will not for fear they should be employed, and set to work." If there is a more serious argument in support of the claim that human language capacity is shared with other primates, then I am unaware of it. In fact, whatever evidence we do have seems to me to support the view that the ability to acquire and use language is a species-specific human capacity, that there are very deep and restrictive principles that determine the nature of human language and are rooted in the specific character of the human mind. Obviously arguments bearing on this hypothesis cannot be definitive or conclusive, but it appears to me, nevertheless, that even in the present stage of our knowledge, the evidence is not inconsiderable.

There are any number of questions that might lead one to undertake a study of language. Personally, I am primarily intrigued by the possibility of learning something, from the study of language, that will bring to light inherent properties of the human mind. We cannot now say anything particularly informative about the normal creative use of language, in itself. But I think that we are slowly coming to understand the mechanisms that make possible this creative use of language, the use of language as an instrument of free thought and expression. Speaking again from a personal point of view, to me the most interesting aspects of contemporary work in grammar are the attempts to formulate principles of organization of language which, it is proposed, are universal reflections of properties of mind; and the attempt to show that on this assumption, certain facts about particular languages can be explained. Viewed in this way, linguistics is simply a part of human psychology: the field that seeks to determine the nature of human mental capacities and to study how these capacities are put to work. Many psychologists would reject a characterization of their discipline in these terms, but this reaction seems to me to indicate a serious inadequacy in their conception of psychology, rather than a defect in the formulation itself. In any event, it seems to me that these are proper terms in which to set the goals of contemporary linguistics, and to discuss its achievements and its failings.

I think it is now possible to make some fairly definite proposals about the organization of human language and to put them to empirical test. The theory of transformational-generative grammar, as it is evolving along diverse and sometimes conflicting paths, has put forth such proposals; and there has been, in the past few years, some very productive and suggestive work that attempts to refine and reconstruct these formulations of the processes and structures that underlie human language.

The theory of grammar is concerned with the question, What is the nature of a person's knowledge of his language, the knowledge that enables him to make use of language in the normal, creative fashion? A person who knows a language has mastered a system of rules that assigns sound and meaning in a definite way for an infinite class of possible sentences. Each language thus consists (in part) of a certain pairing of sound and meaning over an infinite domain. Of course, the person who knows the language has no consciousness of having mastered these rules or of putting them to use, nor is there any reason to suppose that this knowledge of the rules of language can be brought to consciousness. Through introspection, a person may accumulate various kinds of evidence about the sound-meaning relation determined by the rules of the language that he has mastered; there is no reason to suppose that he can go much beyond this surface level of data so as to discover, through introspection, the underlying rules and principles that determine the relation of sound and meaning. Rather, to discover these rules and principles is a typical problem of science. We have a collection of data regarding sound-meaning correspondence, the form and interpretation of linguistic expressions, in various languages. We try to determine, for each language, a system of rules that will account for such data. More deeply, we try to establish the principles that govern the formation of such systems of rules for any human language.

The system of rules that specifies the sound-meaning relation for a given language can be called the "grammar"—or, to use a more technical term, the "generative grammar"—of this language. To say that a grammar "generates" a certain set of structures is simply to say that it specifies this set in a precise way. In this sense, we may say that the grammar of a language generates an infinite set of "structural descriptions," each structural description being an abstract object of some sort that determines a particular sound, a particular meaning, and whatever formal properties and configurations serve to mediate the relation between sound and meaning. For example, the grammar of English generates structural descriptions for the sentences I am now speaking; or, to take a simpler case for purposes of illustration, the grammar of English would generate a structural description for each of these sentences:

1. John is certain that Bill will leave.
2. John is certain to leave.

Each of us has mastered and internally represented a system of grammar that assigns structural descriptions to these sentences; we use this knowledge, totally without awareness or even the possibility of awareness, in producing these sentences of understanding them when they are produced by others. The structural descriptions include a phonetic representation of the sentences and a specification of their meaning. In the case of the cited examples 1 and 2, the structural descriptions must convey roughly the following information: They must indicate that in the case of 1, a given psychological state (namely, being certain that Bill will leave) is attributed to John; whereas in the case of 2, a given logical property (namely, the property of being certain) is attributed to the proposition that John will leave. Despite the superficial similarity of form of these two sentences, the structural descriptions generated by the grammar must indicate that their meanings are very different: One attributes a psychological state to John, the other attributes a logical property to an abstract proposition. The second sentence might be paraphrased in a very different form: 3. That John will leave is certain.

For the first there is no such paraphrase. In the paraphrase 3 the "logical form" of 2 **is expressed more** directly, one might say. The grammatical relations in 2 and 3 are very similar, despite the difference of surface form; the grammatical relations in 1 and 2 are very different, despite the similarity of surface form. Such facts as these provide the starting point for an investigation of the grammatical structure of English—and more generally, for the investigation of the general properties of human language.

To carry the discussion of properties of language further, let me introduce the term "surface structure" to refer to a representation of the phrases that constitute a linguistic expression and the categories to which these phrases belong. In sentence 1, the phrases of the surface structure include: "that Bill will leave," which is a full proposition; the noun phrases "Bill" and "John"; the verb phrases "will leave" and "is certain that Bill will leave," and so on. In sentence 2, the surface structure includes the verb phrases "to leave" and "is certain to leave"; but the surface structure of 2 includes no proposition of the form "John will leave," even though this proposition expresses part of the meaning of "John is certain to leave", and appears as a phrase in the surface structure of its paraphrase, "that John will leave is certain." In this sense, surface structure does not necessarily provide an accurate indication of the structures and relations that determine the meaning of a sentence, in the case of sentence 2, "John is certain to leave", the surface structure fails to indicate that the proposition "John will leave" expresses a part of the meaning of the sentence—although in the other two examples that I gave the structure comes rather close to indicating the semantically significant relations.

Continuing, let me introduce the further technical term "deep structure" to refer to a representation of the phrases that play a more central role in the semantic interpretation of a sentence. In the case of 1 and 3, the deep structure might not be very different from the surface structure. In the case of 2, the deep structure will be very different from the surface structure, in that it will include some such proposition as "John will leave" and the predicate "is certain" applied to this proposition, though nothing of the sort appears in the surface structure. In general, apart from the simplest examples, the surface structures of sentences are very different from their deep structure.

The grammar of English will generate, for each sentence, a deep structure, and will contain rules showing how this deep structure is related to a surface structure. The rules expressing the relation of deep and surface structure are called "grammatical transformations." Hence the term "transformational-generative grammar." In addition to rules defining deep structures, surface structures, and the relation between them, the grammar of English contains further rules that relate these "syntactic objects" (namely, paired deep and surface structures) to phonetic representations on the one hand, and to representations of meaning on the other. A person who has acquired knowledge of English has internalized these rules and makes use of them when he understands or produces the sentences just given as examples, and an indefinite range of others.

Evidence in support of this approach is provided by the observation that interesting properties of English sentences can be explained directly in terms of the deep structures assigned to them. Thus consider once again the two sentences 1 ("John is certain that Bill will leave") and 2 ("John is certain to leave"). Recall that in the case of the first, the deep structure and surface structure are virtually identical, whereas in the case of the second, they are very different. Observe also that in the case of the first, there is a corresponding nominal phrase, namely, "John's certainty that Bill will leave (surprised me)"; but in the case of the second, there is no corresponding nominal phrase. We cannot say "John's certainty to leave surprised me." The latter nominal phrase is intelligible, I suppose, but it is not well formed in English. The speaker of English can easily make himself aware of this fact, though the reason for it will very likely escape him. This fact is a special case of a very general property of English: Namely, nominal phrases exist corresponding to sentences that are very close in surface form to deep structure, but not corresponding to such sentences that are remote in surface form from deep structure. Thus "John is certain that Bill will leave," being close in surface form to its deep structure, corresponds to the nominal phrase "John's certainty that Bill will leave"; but there is no such phrase as "John's certainty to leave" corresponding to "John is certain to leave," which is remote from its deep structure.

The notions of "closeness" and "remoteness" can be made quite precise. When we have made them precise, we have an explanation for the fact that nominalizations exist in certain cases but not in others—though were they to exist in these other cases, they would often be perfectly intelligible. The explanation turns on the notion of deep structure: In effect, it states that nominalizations must reflect the properties of deep structure. There are many examples that illustrate this phenomenon. What is important is the evidence it provides in support of the view that deep structures which are often quite abstract exist and play a central role in the grammatical processes that we use in producing and interpreting sentences. Such facts, then, support the hypothesis that deep structures of the sort postulated in transformational-generative grammar are real mental structures. These deep structures, along with the transformation rules that relate them to surface structure and the rules relating deep and surface structures to representations of sound and meaning, are the rules that have been mastered by the person who has learned a language. They constitute his knowledge of the language; they are put to use when he speaks and understands.

The examples I have given so far illustrate the role of deep structure in determining meaning, and show that even in very simple cases, the deep structure may be remote from the surface form. There is a great deal of evidence indicating that the phonetic form of a sentence is determined by its surface structure, by principles of an extremely interesting and intricate sort that I will not try to discuss here. From such evidence it is fair to conclude that surface structure determines phonetic form, and that the grammatical relations represented in deep structure are those that determine meaning. Furthermore, as already noted, there are certain grammatical processes, such as the process of nominalization, that can be stated only in terms of abstract deep structures.

The situation is complicated, however, by the fact that surface structure also plays a role in determining semantic interpretation. The study of this question is one of the most controversial aspects of current work, and, in my opinion, likely to be one of the most fruitful. As an illustration, consider some of the properties of the present perfect aspect in English—for example, such sentences as "John has lived in Princeton." An interesting and rarely noted feature of this aspect is that in such cases it carries the presupposition that the subject is alive. Thus it is proper for me to say "I have lived in Princeton" but, knowing that Einstein is dead, I would not say "Einstein has lived in Princeton." Rather, I would say "Einstein lived in Princeton." (As always, there are complications, but this is accurate as a first approximation.) But now consider active and passive forms with present perfect aspect. Knowing that John is dead and Bill alive, I can say "Bill has often been visited by John," but not "John has often visited Bill"; rather, "John often visited Bill." I can say "I have been taught physics by Einstein" but not "Einstein has taught me physics"; rather, "Einstein taught me physics." In general, active and passive are synonymous and have essentially the same deep structures. But in these cases, active and passive forms differ in the presuppositions they express; put simply, the presupposition is that the person denoted by the surface subject is alive. In this respect, the surface structure contributes to the meaning of the sentence in that it is relevant to determining what is presupposed in the use of a sentence.

Carrying the matter further, observe that the situation is different when we have a conjoined subject. Thus given that Hilary is alive and Marco Polo dead, it is proper to say "Hilary has climbed Mt. Everest" but not "Marco Polo has climbed Mt. Everest"; rather, again, "Marco Polo climbed Mt. Everest." (Again, I overlook certain subtleties and complications.) But now consider the sentence "Marco Polo and Hilary (among others) have climbed Mt. Everest." In this case, there is no expressed presupposition that Marco Polo is alive, as there is none in the passive "Mt. Everest has been climbed by Marco polo (among others)."

Notice further that the situation changes considerably when we shift from the normal intonation, as in the cases I have just given, to an intonation contour that contains a contrastive or expressive stress. The effect of such intonation on presupposition is fairly complex. Let me illustrate with a simple case. Consider the sentence "The Yankees played the Red Sox in Boston." With normal intonation, the point of main stress and highest pitch is the word "Boston" and the sentence might be an answer to such questions as "where did the Yankees play the Red Sox?" ("in Boston"); "what did the Yankees do?" ("they played the Red Sox in Boston"); "what happened?" ("the Yankees played the Red Sox in Boston"). But suppose that contrastive stress is placed on "Red Sox," so that we have "The Yankees played the RED SOX in Boston." Now, the sentence can be the answer only to "Who did the Yankees play in Boston?" Note that the sentence presupposes that the Yankees played someone in Boston; if there was no game at all, it is improper, not just false, to say "The Yankees played the RED SOX in Boston." In contrast, if there was no game at all, it is false, but not improper, to say "The Yankees played the Red Sox in Boston," with normal intonation. Thus contrastive stress carries a presupposition in a sense in which normal intonation does not, though normal intonation also carries a presupposition in another sense; thus it would be improper to answer the question "Who played the Red Sox in Boston?" with "The Yankees played the Red Sox in Boston" (normal intonation). The same property of contrastive stress is shown by the so-called cleft sentence construction. Thus the sentence "It was the YANKEES who played the Red Sox in

Boston" has primary stress on "Yankees," and presupposes that someone played the Red Sox in Boston. The sentence is improper, not just false, if there was no game at all. These phenomena have generally been overlooked when the semantic role of contrastive stress has been noted.

To further illustrate the role of surface structure in determining meaning, consider such sentences as this: "John is tall for a pygmy." This sentence presupposes that John is a pygmy, and that pygmies tend to be short; hence given our knowledge of the Watusi, it would be anomalous to say "John is tall for a Watusi." On the other hand, consider what happens when we insert the word "even" in the sentence. Inserting it before "John" we derive: "Even John is tall for a pygmy." Again, the presupposition is that John is a pygmy and that pygmies are short. But consider: "John is tall even for a pygmy." This presupposes that pygmies are tall; it is therefore a strange sentence, given our knowledge of the facts, as compared, say, to "John is tall even for a Watusi," which is quite all right. The point is that the position of "even" in the sentence "John is tall for a pygmy" determines the presupposition with respect to the average height of pygmies.

But the placement of the word "even" is a matter of surface structure. We can see this from the fact that the word "even" can appear in association with phrases that do not have any representation at the level of deep structure: Consider, for example, the sentence "John isn't certain to leave at 10; in fact, he isn't even certain to leave at all." Here, the word "even" is associated with "certain to leave," a phrase which, as noted earlier, does not appear at the level of deep structure. Hence in this case as well properties of surface structure play a role in determining what is presupposed by a certain sentence.

The role of surface structure in determining meaning is illustrated once again by the phenomenon of pronominalization. Thus if I say "Each of the men hates his brothers," the word "his" may refer to one of the men; but if I say "The men each hate his brothers," the word "his" must refer to some other person, not otherwise referred to in the sentence. However, the evidence is strong that "each of the men" and "the men each" derive from the same deep structure. Similarly, it has been noted that placement of stress plays an important role in determining pronominal reference. Consider the following discourse: "John washed the car; I was AFRAID someone else would do it." The sentence implies that I hoped that John would wash the car, and I'm happy that he did. But now consider the following:"John washed the car; I was AFRAID someone else would do it." With stress on "afraid," the sentence implies that I hoped that John would not wash the car. The reference of "someone else" is different in the two cases. There are many other examples that illustrate the role of surface structure in determining pronominal reference.

To complicate matters still further, deep structure too plays a role in determining pronominal reference. Thus consider the sentence "John appeared to Bill to like him." Here, the pronoun "him" may refer to Bill but not John. Compare "John appealed to Bill to like him." Here, the pronoun may refer to John but not Bill. Thus we can say "John appeared to Mary to like him," but not "John appeared to Mary to like him," where "him" refers to "John"; on the other hand, we can say "John appealed to Mary to like her," but not "John appealed to Mary to like her," where "her" refers to Mary. Similarly, in "John appealed to Bill to like himself," the reflexive refers to Bill; but in "John appeared to Bill to like himself," it refers to John. These sentences are approximately the same in surface structure; it is the differences in deep structure that determine the pronominal reference.

Hence pronominal reference depends on both deep and surface structure. A person who knows English has mastered a system of rules which make use of properties of deep and surface structure in determining pronominal reference. Again, he cannot discover these rules by introspection. In fact, these rules are still unknown, though some of their properties are clear.

To summarize: The generative grammar of a language specifies an infinite set of structure descriptions, each of which contains a deep structure, a surface structure, a phonetic representation, a semantic representation, and other formal structures. The rules relating deep and surface structure—the so-called "grammatical transformations"—have been investigated in some detail, and are fairly well understood. The rules that relate surface structure and phonetic representation are also reasonably well understood (though I do not want to imply that the matter is beyond dispute; far from it). It seems that both deep and surface structure enter into the determination of meaning. Deep structure provides the grammatical relations of predication, modification, and so on, that enter into the determination of meaning. On the other hand, it appears that matters of focus and presupposition, topic and comment, the scope of logical elements, and pronominal reference are determined, in part at least, by surface structure. The rules that relate syntactic structures to representations of meaning are not at all well understood. In fact, the notion "representation of meaning" or "semantic representation" is itself highly controversial. It is not clear at all. that it is possible to distinguish sharply between the contribution of grammar to the determination of meaning, and the contribution of so-called "pragmatic considerations," questions of fact and belief and context of utterance. It is perhaps worth mentioning that rather similar questions can be raised about the notion "phonetic representation." Although the latter is one of the best established and least controversial notions of linguistic theory, we can, nevertheless, raise the question whether or not it is a legitimate abstraction, whether a deeper understanding of the use of language might not show that factors that go beyond grammatical structure enter into the determination of perceptual representations and physical form in an inextricable fashion, and cannot be separated, without distortion, from the formal—rules that interpret surface structure as phonetic form.

So far, the study of language has progressed on the basis of a certain abstraction: Namely, we abstract away from conditions of use of language and consider formal structure and the formal operations that relate them. Among these formal structures are those of syntax, namely, deep and surface structures; and also the phonetic and semantic representations, which we take to be certain formal objects related to syntactic structures by certain well-defined operations. This process of abstraction is in no way illegitimate, but one must understand that it expresses a point of view, a hypothesis about the nature of mind, that is not a priori obvious. It expresses the working hypothesis that we can proceed with the study of "knowledge of language"—what is often called "linguistic competence"—in abstraction from the problems of how language is used. The working hypothesis is justified by the success that is achieved when it is adopted. A great deal has been learned about the mechanisms of language, and, I would say, about the nature of mind, on. the basis of this hypothesis. But we must be aware that in part, at least, this approach to language is forced upon us by the fact that our concepts fail us when we try to study the use of language. We are reduced to platitudes or to observations which, though perhaps quite interesting, do not lend themselves to systematic study by means of the intellectual tools presently available to us. On the other hand, we can bring to the study of formal structures and their relations a wealth of experience and understanding. It may be that at this point we are facing a problem of conflict between significance and feasibility, a conflict of the sort that I mentioned earlier in this paper. I do not believe that this is the case, but it is possible. I feel fairly confident that the abstraction to the study of formal mechanisms of language is appropriate; my confidence arises from the fact that many quite elegant results have been achieved on the basis of this abstraction. Still, caution is in order. It may be that the next great advance in the study of language will require the forging of new intellectual tools that permit us to bring into consideration a variety of questions that have been cast into the waste-bin of "pragmatics," so that we could proceed to study questions that we know how to formulate in an intelligible fashion.

As noted, I think that the abstraction to linguistic competence is legitimate. To go further, I believe that the inability of modern psychology to come to grips with the problems of human intelligence is in part, at least, a result of its unwillingness to undertake the study of abstract structures and mechanisms of mind. Notice that the approach to linguistic structure that I have been outlining has a highly traditional flavor to it. I think it is no distortion to say that this approach makes precise a point of view that was inherent in the very important work of the 17th- and 18th-century universal grammarians, and that was developed, in various ways, in rationalist and romantic philosophy of language and mind. The approach deviates in many ways from a more modern, and in my opinion quite erroneous, conception that knowledge of language can be accounted for as a system of habits, or in terms of stimulus-response connections, principles of "analogy" and "generalization," and other notions that have been explored in 20th-century linguistics and psychology, and that develop from traditional empiricist speculation. The fatal inadequacy of all such approaches, I believe, results from their unwillingness to undertake the abstract study of linguistic competence. Had the physical sciences limited themselves by similar methodological strictures, we would still be in the era of Babylonian astronomy.

One traditional concept that has reemerged in current work is that of "universal grammar," and I want to conclude by saying just a word about this topic. There are two kinds of evidence suggesting that deep-seated formal conditions are satisfied by the grammars of all languages. The first kind of evidence is provided by the study of a wide range of languages. In attempting to construct generative grammars for languages of widely varied kinds, investigators have repeatedly been led to rather similar assumptions as to the form and organization of such generative systems. But a more persuasive kind of evidence bearing on universal grammar is provided by the study of a single language. It may at first seem paradoxical that the intensive study of a single language should provide evidence regarding universal grammar, but a little thought about the matter shows that this is a very natural consequence.

To see this, consider the problem of determining the mental capacities that make language acquisition possible. If the study of grammar—of linguistic competence—involves an abstraction from language use, then the study of the mental capacities that make acquisition of grammar possible involves a further, second-order abstraction. I see no fault in this. We may formulate the problem of determining the intrinsic characteristics of a device of unknown properties that accepts as "input" the kind of data available to the child learning his first language, and produces as "output" the generative grammar of that language. The "output," in this case, is the internally represented grammar, mastery of which constitutes knowledge of the language. If we undertake to study the intrinsic structure of a language-acquisition device without dogma or prejudice, we arrive at conclusions which, though of course only tentative, still seem to me both significant and reasonably well-founded. We must attribute to this device enough structure so that the grammar can be constructed within the empirically given constraints of time and available data, and we must meet the empirical condition that different speakers of the same language, with somewhat different experience and training, nevertheless acquire grammars that are remarkably similar, as we can determine from the ease with which they communicate and the correspondences among them in the interpretation of new sentences. It is immediately obvious that the data available to the child is quite limited—the number of seconds in his lifetime is trivially small as compared with the range of sentences that he can immediately understand and can produce in the appropriate manner. Having some knowledge of the characteristics of the acquired grammars and the limitations on the available data, we can formulate quite reasonable and fairly strong empirical hypotheses regarding the internal structure of the language-acquisition device that constructs the postulated grammars from the given data. When we study this question in detail, we are, I believe, led to attribute to the device a very rich system of constraints on the form of a possible grammar; otherwise, it is impossible to explain how children come to construct grammars of the kind that seem empirically adequate under the given conditions of time and access to data. But if we assume, furthermore, that children are not genetically predisposed to learn one rather than another language, then the conclusions we reach regarding the language-acquisition device are conclusions regarding universal grammar. These conclusions can be falsified by showing that they fail to account for the construction of grammars of other language, for example. And these conclusions are further verified if they serve to explain facts about other languages. This line of argument seems to me very reasonable in a general way, and when pursued in detail it leads us to strong empirical hypotheses concerning universal grammar, even from the study of a particular language.

I have discussed an approach to the study of language that takes this study to be a branch of theoretical human psychology. Its goal is to exhibit and clarify the mental capacities that make it possible for a human to learn and use a language. As far as we know, these capacities are unique to man, and have no significant analogue in any other organism. If the conclusions of this research are anywhere near correct, then humans must be endowed with a very rich and explicit set of mental attributes that determine a specific form of language on the basis of very slight and rather degenerate data. Furthermore, they make use of the mentally represented language in a highly creative way, constrained by its rules but free to express new thoughts that relate to past experience or present sensations only in a remote and abstract fashion. If this is correct, there is no hope in the study of the "control" of human behavior by stimulus conditions, schedules of reinforcement, establishment of habit structures, patterns of behavior, and so on. Of course, one can design a restricted environment in which such control and such patterns can be demonstrated, but there is no reason to suppose that any more is learned about the range of human potentialities by such methods than would be learned by observing humans in a prison or an army—or in many a schoolroom. The essential properties of the human mind will always escape such investigation. And if I can be pardoned a final "nonprofessional" comment, I am very happy with this outcome.

Notes

1. Rene Descartes (1596—1650) was renowned as a French mathematician, philosopher and physiologist. Though the great philosophical distinction between mind and body in western thought can be traced to the Greeks, it is to the seminal work of Descartes that we owe the first systematic account of the mind and body relationship.

2. Antoine Le Grand(1629—1699)was a philosopher and catholic theologian who played an important role in propagating the Cartesian philosophy in England during the latter half of the seventeenth century. It is not clear how Le Grand came to Cartesianism, but the first evidence of his adoption of the new philosophy was in his *Institutio Philosophiae*, published in London in 1672. He is noted for having given Descartes' work a scholastic form so that it would be accepted in the schools.

3. Chomsky discusses this matter in some detail in "*Deep Structure and Semantic Interpretation*," in R. Jakobson, and S. Kawamoto,eds. ,*Studies in General and Oriental Linguistics*, commemorative volume for Shiro Hattori, TEC Corporation for Language and Education Research,Tokyo,1970.

4. The examples that follow are due to Ray Dougherty, Adrian Akmajian,and Ray jackendoff. See Chomsky's article in Jakobson and kawamoto, eds. , Studies in General and Oriental Linguistics, for references.

Words and Expressions

analogous [əˈnæləgəs]	adj.	相似的,可比拟的;
analogous to	adj.	类似于,类同于
novelty [ˈnɒvlti]	n.	新奇;新奇的事物;新颖小巧价廉的物品
configuration [kənˌfɪgəˈreɪʃn]	n.	布局,构造;配置;结构
novel [ˈnɒvl]	adj.	新奇的;异常的
analogies [əˈnælədʒiz]	n.	类似(analogy 复数),相似,类推,类推法
appraisal [əˈpreɪzəl]	n.	评价,估量;鉴定

objectionable		adj.	令人不快(反感,讨厌)的,有异议的,反对的
inculcation	[ˌinkʌlˈkeiʃn]	n.	灌输,谆谆教
arithmetic	[əˈriθmətik]	n.	算术,计算;算法
incorporated	[inˈkɔːpəreit]	vt./vi.	包含;吸收,合并,混合,结合,并入,编入
feasibility	[ˌfiːzəbiləti]	n.	可行性;可能性;现实性
descartes	[deˈkɑrt]	n.	笛卡尔(法国哲学家、数学家,1596—1690)
authoritarian	[ɔːˌθɔriˈteəriən]	adj.	权力主义的;独裁主义的
enlightenment	[inˈlatnmənt]	n.	启迪,启发;启蒙运动
metaphorically	[ˌmetəˈfɒrikli]	adv.	比喻地,隐喻地
sensual	[ˈsenʃuəl]	adj.	感觉的;肉体上的;色情的,世俗的
recursive	[riˈkəːsiv]	adj.	回归的,递归的 循环的
Indies	[ˈindiz]	n.	印度地方,东西印度群
baboon	[bæˈbuːn]	n.	狒狒
be imbued with		v.	充满;满怀
definitive	[diˈfinitiv]	adj.	最后的;确定的,决定性的
conclusive	[kənˈkluːsiv]	adj.	决定性的;令人信服的;确凿的;结论性的
inconsiderable	[ˌinkənˈsidrəbl]	adj.	价值低的;不值得考虑的;无足轻重的
intrigued	[inˈtriːgd]	adj.	好奇的,被迷住了的
informative	[inˈfɔːmətiv]	adj.	提供信息的;增长见闻的
contemporary	[kənˈtemprəri]	adj.	当代的,现代的;同时代的
inadequacy	[inˈædikwəsi]	n.	不充分,不适当;不完全
empirical	[imˈpirikəl]	adj.	凭经验的,经验主义的,以观察或实验为依据的
introspection	[ˌintrəˈspekʃn]	n.	反省,内省
specification	[ˌspesifiˈkeiʃn]	n.	规格;详述;说明书
be attributed to			把…归于;缘于
given	[ˈgivn]	adj.	指定的,确定的;假设的
property	[ˈprɒpəti]	n.	特性,属性
proposition	[ˌprɒpəˈziʃn]	n.	命题;建议;主张.

paraphrase	['pærə,freizd]	v.	释义，意译
nominal	['nɒminl]	adj.	名义上的；n. 名词性词
correspond to			相当于…，与…相一致
postulate	['pɒstjulet]	vt.	假定；视…为理所当然
representation	[,reprizen'teiʃn]	n.	表现；陈述；
interpretation	[in,tə:pri'teiʃn]	n.	理解，演绎；解释，说明
contour	['kɒntʊə(r)]	n.	外形，轮廓，等高线
Red Sox			红袜队
pygmy	['pigmi]	n.	俾格米人；特别矮小的
watusi	[wɑ:'tu:si]		瓦图西，非洲部落
anomalous	[ə'nɑ:mələs]	adj.	不规则的；不协调的；不恰当
pronominal	[prəʊ'nɒminl]	adj.	代词的；代名词的
appear	[ə'piə]	v.	出现，显出
predication	[,predi'keiʃən]	n.	论断，述谓；述谓结构
modification	[,mɒdifi'keiʃn]	n.	修改，变更，改良，[语]修饰
controversial	[,kɒntrə'və:ʃl]	adj.	有争议的
legitimate	[li'dʒitimət]	adj.	合法的，合理的；真正的，真实的
perceptual	[pə'septʃuəl]	adj.	知觉的，有知觉的；感性
inextricable	[in'ekstrikəbəl]	adj.	无法摆脱的，解不开的
platitude	['plætitju:d]	n.	平常的话，老生常谈，陈词滥调
intelligible	[in'telidʒəbl]	adj.	可理解的，明白易懂的，清楚的
rationalist	['ræʃnəlist]	n.	唯理论者，理性主义者
deviates	['divi,et]	v.	脱离，越轨，违背，误入歧途
erroneous	[i'rəuniəs]	adj.	错误的；不正确的
accounted for			说明（原因、理由等），导致，引起，对…负责
analogy	[ə'nælədʒi]	n.	类似，相似；比拟，类比；类推
empiricist	[im'pirisist]	n.	经验主义者，经验论者
stricture	['striktʃə]	n.	苛评；指责；紧束；约束
babylonian	[,bæbi'ləunjən]	n.	巴比伦（人，国）
astronomy	[ə'strɒnəmi]	n.	天文学
formulate	['fɔ:mjuleit]	vt.	构想出，规划
intrinsic	[in'trinsik]	adj.	固有的，内在的，本质的
device without dogma	['dɒgmə]	n.	教条，信条

empirically	[im'pirikli]	adv.	以经验为主地
trivially	['triviəli]	adv.	琐细地；很一般的；显而易见地
postulate	['pɔstʃəˌleit]	v.	假定，假设
analogue	['ænəˌlɔg]	n.	相似物；相似的情况
endow	[in'dau]	vt.	捐赠，赋予，赐予
sensation	[sen'seiʃn]	n.	感觉；直觉；知觉

Questions for Discussion and Review

1. explain in your own words what Chomsky believes to be "the core problem of human language…."

2. Do you agree with Chomsky that "This creative aspect of normal language use is one fundamental factor that distinguish human language from any know system of animal communication"? Defend your ansmer.

3. Explain why Chomsky insists on the importance, especially in schools at all levels, of the recognition of the limitations of our understanding of linguistic and intellectual creativity.

4. explain Chomsky's assertion that "Viewed in this way, linguistics is simply a part of human psychology…"

Unit 12

Semantics and Semantic Theory

> The present selection is from the book *Formal Semantics: An Introduction* written by Ronnie Cann. After a brief mention of the distinction between general linguistic semantics and formal semantics, and Montague's contribution to the formal study of natural language, the author moves to the discussion of some fundamental properties of semantic theory, making clear the principles of semantic study of language. In the sketch here, many of the research areas in semantics and their relations to the original research or researchers are mentioned, which is certainly of help for follow-up studies.
>
> Precision in description is one of the advantages of the formal approach to the study of language, which is characteristic of Natural Language Processing (NLP). The advance or breakthrough in this area of study will therefore be of significance to the research in NLP.

In its broadest sense, semantics is the study of meaning and linguistic semantics is the study of meaning as expressed by the words, phrases and sentences of human languages. It is, however, more usual within linguistics to interpret the term more narrowly, as concerning the study of those aspects of meaning encoded in linguistic expressions that are independent of their use on particular occasions by particular individuals within a particular speech community. In other words, semantics is the study of meaning abstracted away from those aspects that are derived from the intentions of speakers, their psychological states and the socio-cultural aspects of the context in which their utterances are made. A further narrowing of the term is also commonly made in separating the study of semantics from that of pragmatics. Unfortunately, the nature of the object of inquiry of the discipline (what constitutes semantic meaning, as opposed to pragmatic meaning) and the domain of the inquiry (what aspects of meaning should be addressed by the discipline) remain difficult and controversial questions. There are, however, three central aspects of the meaning of linguistic expressions that are currently accepted by most semanticists as forming the core concern of linguistic semantics. These central concerns of semantic theory, adapted from Kempson(1977:4), are stated in (1) and may be adopted as criteria for ascertaining the adequacy of semantic theories which apply in addition to the general conditions on scientific theories of falsifiability and rigour.

(1) A semantic theory must:

a. capture for any language the nature of the meaning of words, phrases and sentences and explain the nature of the relation between them;

b. be able to predict the ambiguities in the expressions of a language;

c. characterise and explain the systematic meaning relations between the words, the phrases and the sentences of a language.

One may add to these the condition that a semantic theory should provide an account of the relation between linguistic expressions and what may be called 'things in the world'. In other words, it is a primary concern of a semantic theory to explain how human beings can use their language to convey information about the external world. We may thus require a semantic theory to conform also to the criterion of adequacy in (2)

(2) A semantic theory must provide an account of the relation between linguistic expressions and the things that they can be used to talk about.

There are many other aspects of meaning that can be included in the domain of linguistic semantics, but a theory conforming to the four criteria in (1) and (2) will cover the main ground of the discipline and provide a firm basis for further research. In this book, we will be looking at a particular theory of semantics that goes a long way towards satisfying these criteria and that has been very influential in linguistic semantics over the last two decades. This theory is a formal theory of semantics and is distinguished from general linguistic semantics by its greater use of mathematical techniques and reliance on logical precision. This is not to say that formal semantics and general linguistic semantics are completely separate disciplines. It sometimes appears that these two approaches to the semantics of natural languages are mutually incompatible, but this is not obviously true. The former draws heavily on the long tradition of research in the latter which in turn benefits from the greater precision of the former. Both approaches enable us to understand more about meaning and greater integration between them would doubtless bring greater benefits to the discipline.

Formal semantics itself was devised as a means of providing a precise interpretation for formal languages, i. e. the logical and mathematical languages that are opposed to natural languages that are spoken or written as the native languages of human beings. Many logicians considered it to be impossible to apply the same rigour to the semantics of human languages, because of their supposedly inexact syntax, their vagueness and their ambiguity. In the late nineteen-sixties, however, the philosopher Richard Montague asserted that it was possible to use the same techniques in analyzing the meanings of sentences in English. In three articles, *English as a formal language*, *Universal grammar* and *The proper treatment of quantification in English*, all published or presented in 1970, Montague gave arguments for his hypothesis that:

There is in my opinion no important theoretical difference between natural languages and the artificial languages of logicians; indeed, I consider it possible to comprehend the syntax and semantics of both kinds of language within a single, natural and mathematically precise theory.

Montague (1974:222)

Throughout the nineteen-seventies, after his tragic death in 1971, Montague's work had a radical effect on the study of semantics in linguistics. Indeed, his ideas on the semantics of human languages have become central to the understanding of many of the questions and theories being discussed in linguistic semantics today. Owing to the relatively recent application of the tools of formal semantics to the analysis of natural languages, however, there are many topics in linguistic semantics that have not yet been formally analysed, but it is hoped that ultimately a good deal of linguistic meaning will be amenable to the sort of rigorous treatment envisaged by Montague. It is the exposition of Montague's theory in its now classical form that constitutes the subject matter of this book, but, before the main points of his semantic theory are introduced, the four criteria of adequacy in (1) and (2) above will be discussed in more detail in order to provide a clearer idea of the fundamental issues that underlie the development of the theory in later chapters.

1. Compositionality

A fundamental property that any semantic theory must have is the ability to pair the syntactic expressions of a language with their meanings. In the first condition of adequacy in (1. a), above, this property is characterised as a requirement that a semantic theory account for the nature of the meaning of linguistic ex-

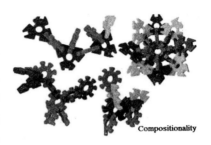
Compositionality

pressions and be able to pair every expression in a language (words, phrases and sentences) with an appropriate meaning. As already mentioned, the characterisation of meaning is a controversial matter, but whatever meanings are taken to be within a theory, it is obvious that there must be some way of associating them with appropriate linguistic expressions. This is not a trivial matter, however, and there are a number of important points that need to be discussed with respect to this property.

In the first place, let us consider more closely what it is that is to be assigned meanings by a semantic theory. Condition (1. a) refers to words and sentences as the carriers of meaning. The term *sentence* here is being used in its abstract sense, common in linguistics, as the largest unit of syntactic description, independently of its realization in spoken or written texts. Like the term sentence, word is also ambiguous in everyday English. Within semantics, the notion of word that is most useful is that of the lexeme which is an abstract grammatical construct that underlies a set of word forms which are recognised as representatives of 'the same word' in different syntactic environments. For example, the word forms *sing*, *sings*, *singing*, *sang* and *sung* are particular inflectional variants of a lexeme which we may represent for the time being as *SING*. It is to lexemes and not to word forms that meanings should be assigned, because while the inflectional properties of the verb SING may vary in different syntactic environments, the sort of action described by the verb remains the same. For this reason, the meaning of words is referred to in this book as lexical meaning, rather than word meaning.

Although it is possible for the meanings assigned to lexemes and sentences to be very different from each other, it is reasonable to expect the meanings of sentences to be related to the meanings of the lexemes underlying the word-forms they contain. It is intuitively implausible for there, to be a language where the relation between the meaning of a sentence and the meanings of its component lexemes is entirely random. While language do contain idiomatic phrases and sentences where lexical and sentential meaning are not transparently related (e. g. *kick the bucket* meaning the same as *DIE* in English), this is never the general situation. If there were no direct relation between lexical and sentential meaning, of course, the meaning of each sentence in a language would have to be listed. Since the number of sentences that make up a language is infinite, this would mean that no human being would be able to determine the meanings of all the sentences of any language owing to the finite resources of the brain. This is absurd, of course, and just as sentences are defined recursively by syntactic rules, taking words (or morphemes) as their basis, so their meanings should also be defined recursively from the meanings ascribed to the lexemes they contain.

Thus, in addition to associating each expression in a language with a meaning, an adequate semantic theory must also be able to explain how the meanings of smaller expressions contribute to the meanings of larger ones that contain them. A theory that derives the meaning of larger expressions from those of smaller ones is said to be compositional. The Principle of Compositionality, given an initial definition in (3), is generally attributed to the German philosopher Gottlob Frege, and is thus sometimes referred to as the Fregean Principle of Compositionality or just the Fregean Principle, although it is unlikely that he ever stated the principle in precisely this way..

(3) The meaning of an expression is a function of the meaning of its parts.

The notion of a function will be discussed in more detail in Chapter 4, but essentially it is an operation that derives a single result given a specified input. Thus, the principle (3) minimally requires that the meaning of a larger expression be uniquely determined from the meanings of its component parts. This cannot be all there is to compositionality, however, since, otherwise, we would expect that sentences containing the same words mean the same thing. This is, of course, not true. The sentence *Jo kicked Chester* does not mean the same as *Chester kicked Jo*. It must also be the case, therefore, that the syntactic structure of an expression is relevant to the derivation of its meaning. Indeed, we may strengthen the principle of compositionality so that, in deriving the meaning of a composite expression, the meaning of its component expressions are combined in some way that refers to the way they are combined by the syntax. This implies that wherever meanings are combined in a particular way to derive the meaning of a composite expression, all other composite expressions of the same sort have their meanings determined in the same way. In other words, the construction of meanings is rule-governed, in the same way that the construction of the well-formed syntactic expressions of a language is rule-governed. For example, whatever rule derives the meaning of the sentence Jo sang from the subject *Jo* and intransitive verb *sang* applies to all declarative sentences derived by combining a subject noun phrase with the appropriate form of an intransitive verb.

Furthermore, it is a general property of human languages that all the sub-expressions of a grammatically well-formed phrase have a role to play in the interpretation of a sentence, even if, on occasions, this role is predictably redundant (as, for example, in double negative constructions in certain dialects of English like *I never did nothing* where the second negative expression merely reinforces the idea of negation introduced by the first). Semantic rules should, therefore, not be allowed to delete meanings during the derivation of the meaning of a composite expression. The effect of this restriction is to make the creation of the meanings of larger expressions monotonic with respect to their component parts where a derivation is said to be monotonic if all properties of previous parts of a derivation are maintained throughout. In other words, once information is introduced into a monotonic derivation, it is not lost thereafter. The initial definition of compositionality in (3) may thus be strengthened to give the statement in (4).

(4) The principle of compositionality: The meaning of an expression is a monotonic function of the meaning of its parts and the way they are put together.

The implications of this interpretation of the principle of compositionality is that meanings should be ascribed not only to lexemes and sentences but also to other syntactic constituents. It is thus generally assumed that meanings should be assigned to all the well-formed constituents of a language, not just to its words (lexemes) and sentences. Indeed, the concept of syntax as a bridge between phonology and semantics, current in many grammatical theories, would seem to require that all constituents be assigned a meaning by the semantics and, furthermore, that (surface) syntactic structure should directly determine how the meanings of sentences are derived. It is common to assume that semantic constituency parallels syntactic constituency and hence that an adequate semantic theory must be able to ascribe appropriate meanings to noun phrases like *the old cat*, *Jo's mother*, *Chester* and verb phrases like *sang*, *kicked the cat*, *ran slowly*, etc., according to their syntactic structure.

One way in which this may be achieved is to adopt the hypothesis that for each syntactic rule of the grammar (or syntactic structure admitted by the grammar) there is a corresponding semantic rule that derives the meaning of the resultant expression (or structure). For example, assuming that there is a rule that defines a sentence in English as consisting of a noun phrase plus a verb phrase, then the adoption of the rule-to-rule hypothesis in (5), together with the principle of compositionality in (4), requires that there be a corresponding semantic rule deriving the meaning of the sentence from the meanings of its immediate constituents, NP and VP.

(5) rule-to-rule hypothesis: for each syntactic rule there is a corresponding semantic rule.

The principle of compositionality in (4) is assumed to be a constraint on semantic theories and, indeed, will be seen to be the primary motivator behind much of the discussion in later chapters. The rule-to-rule hypothesis, on the other hand, is not a necessary requirement of a semantic theory, but a means of achieving compositionality.

2. Meaning relations

Another aspect of meaning that must be accounted for by any semantic theory is the systematic relations that hold between different expressions in a language. According to the condition of adequacy in (1.b), a theory must satisfactorily analyse the intuitions speakers of English have about the semantic relations between lexemes and between sentences. This assumes that expressions in a language which may not be syntactically related may be related semantically and, indeed, such is the case. Consider the sentence in (6). Assuming that the reference of the name Jo and the discourse context are held constant for all the sentences in (6), then the sentences in (6.b) to (6.i) are semantically related to that in (6.a), even though it is not always the case that there is a direct syntactic relation between them.

(6) a. Jo stroked a cat.
    b. A cat was stroked.
    c. There was a cat.
    d. No-one stroked a cat.
    e. There are no such things as cats.
    f. A cat was stroked by Jo.
    g. It was Jo who stroked the cat.
    h. Jo touched a cat.
    i. Jo stroked an animal.

The relation between the sentences in (6.a) and those in (6.b) and (6.c) is one of entailment, as it is intuitively impossible for it to be true that Jo stroked a cat on some occasion without it also being true that a cat was stroked or that there existed a cat to be stroked on that occasion. We may thus define a sentence S1, as entailing a sentence S2 if the latter truly describes a situation whenever the former also does. The negation of an entailment always derives a contradiction and a sentence S1 may be said to contradict a sentence S2 if the former must be false when the latter is true (or vice versa). For, the assertion that JO stroked a cat is contradicted by the non-existence of cats, making (6.e) a contradiction of (6.a). The relation between (6.a) and the sentences in (6.f) and (6.g) is also primarily semantic, although most syntactic theories recognize a syntactic relation between the sentences as well. Using a common term in a technical way, we may say these sentences are paraphrases of each other, since they all have the same core meaning. Another way of putting this is to say that they mutually entail each other. Hence, we may say that a sentence S1 is a paraphrase of a sentence S2 if S1 entails S2 and S2 entails S1. An adequate theory of semantics must, therefore, provide an account of entailment, contradiction and paraphrase that allows one to identify which sentences are entailed by, or contradict or paraphrase, another in a language. Indeed, this concern, along with compositionality, is a major motivation for the theoretical.

Other sorts of implication between sentences are also recognized in general linguistic semantics. Some of these derive from lexical meaning as in (6. h) and (6. i) which are related to (6. a) by virtue of the meanings of the lexemes STROKE and CAT, respectively. It is part of the meaning of the lexeme STROKE that an action of stroking also involves an action of touching, so that (6. a) implies (6. h). Furthermore, it is part of the meaning of CAT that anything that is a cat is also an animal and thus (6. a) implies (6. i). The meaning relations that hold between the lexemes of a language (or between lexemes and larger expressions) are called sense relations and include hyponymy, which holds if the sense of one lexeme includes that of another (e. g. between CAT and ANIMAL); synonymy, where two lexemes have the same sense (e. g. between MERCURY and QUICKSILVER); and oppositeness, where two lexemes have opposing senses (e. g. BIG and SMALL and DEAD and ALIVE). Hence,an adequate theory of semantics must give some account of lexical meaning and,in particular, of the sense relations that hold between lexemes in a particular language.

Other implicational meaning relations hold between sentences in addition to those that have been noted above. One of these is so-called conversational implicature, which is derived not from the conventional meanings of words or sentences,but from principles of discourse and context. For example,in the exchange in (7),the apparent irrelevance of Ethel's reply in (7. b) to Bertie's question in (7. a) leads the latter to infer(7. c). The reason behind inference has to do with Bertie's expectations about Ethel's co-operativeness in providing him with the information he needs. The fact that she has not given a straight answer leads Bertie, not to assume that she is being deliberately unhelpful, but to look for some piece of information that is relevant to his question that is indirectly implied by Ethel's response.

(7)a. Bertie: Is Fiona a good lecturer?

b. Ethel: She has a good line in sweaters...

c. Bertie (thinks-): Fiona is not a gook lecturer.

Another sort of implication between sentences is called presupposition. A sentence is said to presuppose another if its truth and that of its negation both imply that the presupposed sentence is also true. In other words,presupposition deals with aspects of meaning that are assumed to hold of a situation when a sentence is uttered to describe that situation. For example,the use of the definite article in a noun phrase is said to presuppose the existence of something that has the property described by the common noun in the same NP. This is illustrated in (8) where the truth of the sentences in (8. c) and (8. d) is presupposed by that of (8. a) and its negation in (8. b), because of the use of *the* in the subject noun phrase. (8. c) is not implied by (8. e) which replaces *the* in the subject NP by *every* (as shown by the bracketed causal clause which denies the truth of (8. c), apparently without contradiction) and, while (8. f) implies (indeed, according to the discussion above, entails) (8. c), its negation in (8. g) does not.

(8) a. The Duchess of Muckhart terrorised the village.

    b. The Duchess of Muckhart didn't terrorise the village.

    c. There is a Duchess of Muckhart.

    d. There is a village

    e. Every Duchess of Muckhart terrorised the village (because there is no Duchess of Muckhart).

    f. A Duchess of Muckhart terrorized the village.

    g. A Duchess of Muckhart didn't terrorize the village.

It is usually assumed that implicatures such as that in (7) result from principles of conversation and thus form part of the domain of pragmatics rather than semantics. More controversial, however, is the status of presupposition. Whether it should be included in semantic or pragmatic theory is an extremely vexed question, as indeed is the definition and status of the phenomenon itself.

In addition to accounting for these semantic relations, a semantic theory may also be required to provide some account of anomaly in the meaning of expressions in some language. It should, therefore, be able to explain why certain expressions which are syntactically well-formed are unacceptable or deviant from the semantic point of view. For example, the sentence in (9. a) is syntactically well-formed and semantically coherent in English. Those in (9. b) and (9. c), however, are semantically anomalous despite the fact that they have the same syntactic structure as (9. a). Such sentences can, however, be given some sort of non-literal interpretation (although (9. c) is harder to find an interpretation for than (9. b), unlike the completely ill-formed expression in (9. d) which is simply not English. This decline in acceptability from (9. a) to (9. c) and the incoherence of (9. d) should thus be explained by an adequate theory of semantics.

(9) a. Green Wellington boots are very popular now.

    b. Green ideas are very popular now.

c. Green corollaries are very popular now.

d. * very Wellington is boots popular now green.

3. Ambiguity

The third area of meaning that Kempson (1977) suggests must be explained by a semantic theory is ambiguity. A sentence is said to be ambiguous whenever it can be associated with two or more different meanings. Ambiguity can arise in a sentence for a number of reasons; through the ascription of multiple meanings to single words (e. g. (10. a)); through the assignment of different syntactic structures to a sentence (e. g. (l0. b)); or through the use of certain expressions that may have different semantic scope(e. g. (10. C)).

(10)a. Ethel's punch was impressive.

b. The strike was called by radical lecturers and students.

c. Every good politician loves a cause.

The first sort of ambiguity occurs where an expression is associated with two or more unrelated meanings, as in (10. a) where the word punch may be interpreted as a drink or as an action. Lexemes whose word forms have this property are called homonyms and can be subdivided into homophones, where the forms of the lexeme sound the same but may be written differently, e. g. *draft* and *draught* which can both be represented phonemically as/draft/, and homographs, e. g. *lead*, which are written the same, but which are pronounced differently. Some lexemes are both homophones and homographs, like PUNCH. Homonyms can be divided into full homonyms(like, BANK, PUNCH), where all of the lexeme's associated word forms are phonetically or orthographically identical, and partial homonyms (like FIND, FOUND), where just some of its word forms are identical.

Homonymy is often contrasted with polysemy. A polysemous lexeme is one that is interpreted as -having multiple senses that are not entirely distinct, as is the case in the standard examples of homonyms. The classic example of a polyseme in English is the lexeme MOUTH which has different interpretations depending on what sort of entity is described as having a mouth. There are, for example, human mouths, mouths of caves mouths of bottles, mouths of rivers, and so on. In each of these cases, the properties of the entity described by MOUTH are different, but not absolutely different, as each one refers to an opening of some sort. The difference between homonymy and polysemy is one of degree, and precise definitions of these terms are difficult and controversial. As this book is not primarily concerned with lexical meaning, no attempt will be made to differentiate the two notions or to incorporate polysemy within the theory at all. As will be seen in Chapter 2, the approach to homonymy taken here is very simplistic: the sense of homonymous lexemes are simply differentiated formally by the use of superscripts, where necessary. Although an account of polysemy and a better approach to homonymy may be possible within the theory of formal semantics presented in later chapter, these matters are not central to the concerns of this book and an adequate discussion of the issues involved would only serve to increase the size of the book without serving any great purpose. The decision to exclude polysemy from consideration and to take a simplistic view of homonymy is taken on the grounds of expository convenience and readers are again referred to the further reading noted at the end of the chapter.

A more interesting source of ambiguity from the point of view of the formal semanticist is illustrated in (10. b). Here the ambiguity results from the possibility of assigning two or more syntactic structures to a single grammatical string of words. To ascertain the meaning of (10. b), for example, it is necessary to know whether the adjective *radical* modifies the nominal phrase, *lecturers and students*, in which case both the lecturers and the students who called the strike are all radical, or whether it modifies just the noun *lecturers*, in which case the lecturers who called the strike are said to be radical but the political attitude of the students who did so is not specified. These two readings are illustrated in (11) where the labelled bracketings of the agentive noun phrase in (11. b) and (11. d) correspond to the readings indicated in (11. a) and (11. c), respectively.

(11)a. The strike was called by lecturers who are radical and by students.
    b. [NP[N1[N1 radical lecturers] and students]].
    c. The strike was called by lecturers who are radical and by students who are radical.
    d. [NP[N1 radical [N1 lecturers and students]]].

In the above example, what is at issue is the scope of the adjective, *radical*. In (11. a), it modifies, and thus has scope over, the noun lecturers, while in (11. b) its scope is the nominal phrase lecturers and students. Scope is an important concept in semantics and a primary source of ambiguity which involves not only adjectives, but also conjunctions, like *and*, *or*, etc. and quantifiers, like every, all, and some in English. Structural ambiguity of this sort thus has its source in the syntax of a language, but there are other scope ambiguities that do not directly depend on the syntactic structure of a sentence. Such ambiguity usually involves negation (not), quantification (*every*, *some*) and other elements like tense, which do not vary their syntactic position according to the reading of the sentence. For example, the two readings of the sentence in (10. c) can be made clear by those in (12). In (12. a), there is only one cause that every good politician loves, while in (12. b) each politician may love a different cause. The sentence in (10. c), however, is usually only assigned a single surface constituent structure, so that this ambiguity cannot be directly attributed to a syntactic source and is referred to as a semantic scope ambiguity.

(12) a. Every politician loves a cause and that is their own career.
     b. Every good politician loves a cause and each one loves a cause that everyone else loathes.

An adequate semantic theory must thus be able to predict where structural ambiguity is likely to arise in a language and provide a means of differentiating the interpretations of the different structures to an ambiguous sentence by the grammar, where this is relevant. It should also ensure that sentences that have two (or more) syntactic derivations, but only one semantic interpretation, are not assigned more than one meaning. The theory should also provide an account of scope ambiguities where these are not directly reflected in syntactic derivations, and be able to differentiate the scopes of particular expressions independently of the syntax.

4. Denotation

The final criterion of adequacy that is considered here is stated in (2), above, and is the most important for our purposes, since it forms the basis of the semantic theory to be proposed in the rest of this book. This criterion requires a semantic theory to give an account of the relation between linguistic expressions and what they can be used to talk about. Since language can be used to talk about what is outside the linguistic system, it is essential that a semantic theory should be able to associate linguistic expressions with extra-linguistic objects. Language is not used solely to talk about itself, but rather it is most commonly used to convey information about the situations in which human beings find themselves. Since a listener can in general understand the meaning of what is being said by a speaker, meanings must be publicly accessible in some sense. One way that this public accessibility must be realised is in the association of linguistic expressions with publicly identifiable entities and situations. For example, the utterance of a sentence like *The book is on the table* conveys information about two entities, one of which is conventionally called a book in English and one of which is conventionally called a table, and the relation between them. Someone who hears an utterance of this sentence associates it with the situation pictorially represented in (13). Although (13) is itself a representation of an actual (or Possible) situation, it is nonetheless a non—linguistic representation and a theory of semantics should be capable of relating the meaning of the sentence to the Picture and, indeed, to concrete, non — representational situations where there is a (single) book on the table.

The association between the sentence the book is on the table and the situation represented in (13) depends in part on there being, in the situation described, an instance of a thing that is conventionally called a book and one that is conventionally called a table in English. In other words, part of the meaning of the sentence depends on the sorts of extra — linguistic entities that can be referred to by the lexemes BOOK and TABLE. The aspect of the meaning of an expression that concerns its relation to such objects called its denotation and an expression is said to denote particular sorts of extra—linguistic objects. Although this relation has often been called the reference of an expression, this book follows the usage of Lyons (1977) and reserves this latter term for the act of picking out a particular entity denoted by the expression through the utterance of that expression on some occasion. For example, in uttering the sentence *The book is on the table*, a speaker is said to be referring to two Particular, contextually unique, entities. The entities being referred to by the use of the definite noun phrases, the book and the tale, are single elements

The book is on the table

in the class of entities denoted by the lexemes BOOK and TABLE.

Thus, a speaker may use linguistic expressions to refer, but linguistic expressions themselves denote. No more will be said here about the act of reference, and the differences between denotation and reference, but for more details the reader is urged to consult the further reading at the end of this chapter.

Informally, we may think of the denotation as the relation between an expression and a class of various sorts of individuals, events, properties and relations that may be referred to by the use of the expression on some particular occasion. The lexeme BOOK may, therefore, be thought of as denoting the set of all books, TABLE as denoting the set of all tables, while the preposition ON may be thought of as denoting the set of all the pairs of entities of which one is on the other. It is easy to grasp the notion of denotation with respect to lexemes that denote concrete entities like books and tables, but the question arises about whether abstract lexemes like LOVE, KNOWLEDGE or THEOREM or ones denoting fictitious entities like UNICORN or HOBBIT have DENOTATIONS in the same way. The answer is, as might be expected, controversial, but here the position is taken that there is no essential difference between such expressions and those that denote concrete entities. Thus, the noun LOVE is taken to denote a set of entities just like BOOK. The difference between them is that the entities denoted by the former are abstract while those denoted by the latter are concrete. Although the postulation of abstract entities of this sort may cause problems from a philosophical point of view, it does have the advantage of reflecting the fact that the same sorts of linguistic expressions (e. g. nouns) are used in many, if not all, languages to refer to both abstract and concrete things.

In a similar fashion, lexemes describing fictitious entities like hobbits, or entities that are no longer extant like dodos, are also assumed to have a denotation. It is, however, useful to distinguish between the denotations of lexemes that may be used to refer to entities that exist in the real world (including abstract ones) and those that do not. We can do this by making a distinction between two aspects of denotation. Nouns like BOOK may be used to refer to entities in the world, but the entities of which one can truthfully say That is a book all share a certain common property, their "bookness", so to speak. In the same way, the set of entities that are red all have the property of redness and the set of entities that run all have the property of running. In other words, we distinguish between the property denoted by a common noun like BOOK, adjective like RED or intransitive verb like RUN and the entities it can be used to refer to. The former part of the meaning of the lexeme is often referred to as its sense and is opposed to the idea of its reference. However, just as the latter term is used here for a different notion, so too is that of sense which is used solely with respect to the sense relations that hold between the lexemes of a language. The distinction between the different aspects of the meaning of BOOK noted above are treated as a distinction between the intension of an expression and its extension. The former corresponds to the property aspect of common nouns, whereas the latter corresponds to the entities that they may be used to refer to in the world. Thus, the extension of the lexeme BOOK is the set of all books whilst its intension is the property of being a book. Both properties and sets of entities are external to the linguistic system and thus constitute aspects of denotation. This distinction allows a differentiation between entities like books which have existence in the real world and those like unicorns that, presumably, do not. In the latter case, the lexeme UNICORN has no extension in the real world, but it does have an intension, the property of being a unicorn. Thus, we may speak about unicorns and other entities without them needing to exist in the real world. This distinction of intensionality is concerned primarily with the notion of extension and thus be concentrating on the relation between linguistic expressions and existing entities.

It is not necessary to restrict the notion of denotation to lexemes, but it may be extended to all well-formed linguistic expressions. For example, the verb phrase *kicked a cat* may be taken to denote the class of actions involving the kicking of a cat, which is distinct from the class of all actions involving kicking a dog which would be denoted by the expression *kicked a dog*, and so on. The denotations of other expressions, like quantified noun phrases, are less easy to specify informally, but we will see how they can be defined. Those expressions that are not used to pick out external entities in any way may also have denotations. Such expressions are typically described as functional or grammatical expressions, like determiners and conjunctions, as opposed to the content expressions, like nouns, verbs and adjectives. This distinction amongst the syntactic categories of a language is a traditional (and very useful) one and can be reflected in semantics by assigning rather different sorts of denotation to the two sorts of expression. Grammatical expressions are taken to denote logical relations between groups of entities that are denoted by content expressions.

A theory of denotation is thus not a trivial one and any semantic theory that provides an account of this important relation has already achieved a great deal. The terms denotation, denotes, intension and extension are used a great deal. Although denotation most properly refers to the relation between an expression and some entity, event, property or relation, it will also be used below to refer to what an expression denotes (what Lyons (1977) calls the denotatum of an expression). The term is of general importance that include both extensional and intensional denotation. Where reference is being made to what a specific expression denotes, the terms extension or intension are used, depending on which aspect of denotation is relevant. Thus, the extension of an expression is taken below to refer to what the expression extensionally denotes, and similarly for the term intension.

Words and Expressions

sketch [sketʃ]	n.	草图,素描,简述
ascertain [ˌæsə'tein]	vt.	弄清,确定,查明
falsifiability [ˌfɒsifaiə'biliti]		不可靠性,可证伪性
rigour ['rigə(r)]	n.	严厉,苛刻,严密性,严谨
rigorous ['rigərəs]	adj.	严密的,缜密的,严格的
envisage [in'vizidʒ]	vt.	想象,设想;展望
compositionality [cʌmpəziʃə'næliti]		组合性
trivial ['triviəl]	adj.	琐碎的,无价值的,平常的,不重要的
lexeme ['leksiːm]	n.	词位,词素,词汇单位
implausible [im'plɔːzəbl]	adj.	难以置信的,不太可能的
idiomatic [ˌidiə'mætik]	adj.	符合语言习惯,成语的;含有习语的
morpheme ['mɔːfiːm]	n.	形态素,词素
ascribe to		归因于
derivation [ˌderi'veiʃn]	n.	衍生物
intransitive [in'trænsətiv]	adj.	(动词)不及物的
predictably [pri,diktəbli]	adv.	可预见地

with respect to			关于,(至于)谈到
entailment	[enˈteilmənt]	n.	限定继承 蕴涵;蕴含
entail	[inˈteil]	vt.	牵涉;需要;限定继承
by virtue of			由于,因为
hyponymy	[haiˈpɒnimi]	n.	(词义之间的)上下义关系,上下位关系
implicature	[ˈimplikətʃə(r)]	n.	会话含义
vexed question		n.	难题
anomaly	[əˈnɒməli]	n.	异常,反常;异常现象
ascription	[əˈskripʃn]	n.	归属
homophone	[ˈhɒməfəun]	n.	同音异义词
homograph	[ˈhɒməgrɑːf]	n.	同形异义词
orthographically	[ˌɔːθəˈgræfikəl]	adj.	正字法的,拼字正确的
polysemy	[pəˈlisimi]	n.	意义的分歧;多义性;一词多义
homonymy	[hɒˈmɒnimi]	n.	同名;异义;同音异义关系
simplistic	[simˈplistik]	adj.	过分单纯化的;过分简单化的
differentiate		vt. & vi.	区分,区别
denotation	[ˌdiːnəuˈteiʃn]	n.	指示意义;直指
Theorem	[ˈθiərəm]	n.	定理;公理,定律,法则
unicorn	[ˈjuːnikɔːn]	n.	(传说中)独角兽,麒麟

Questions for Discussion and Review

1. What is the difference between generative linguistic semantics and formal semantics?

2. Comment on the relationship among lexical meaning, word meaning and sentential meaning described in the present selection.

3. Recursiveness is a common feature of language use. Illustrate this feature by examples from both English and Chinese.

4. What are the fundamental properties of semantic theory discussed in this selection? Explain each in your own words.

5. Comment on the philosophers' contribution to the study of meaning touched upon in the present selection.

Unit 13

Pragmatics

> Pragmatics, concerned with the study of meaning as communicated by a speaker (or writer) and interpreted by a listener (or reader), has more to do with speaker meaning. Besides, it is the study of contextual meaning, of how more gets communicated than is said, and of the expression of relative distance. Unlike semantics, the study of the relationship between linguistic forms and entities in the world, pragmatics is the study of the relationships between linguistic forms and the users of those forms. In the following excerpt from her book How Language Works, Madelon E. Heatherington discusses the various types of illocutionary force, or the communicative intent of the speaker; conversational principles, or expectations that are shared by all participants in a conversation; and presuppositions, the surprisingly large number of assumptions made by all speakers about what their listeners know. It is still not clear how information about such matters should be included in a grammar, but it is clear that pragmatics must be dealt with if our description of language use is to be complete.

Pragmatics deals with particular utterance in particular situations and is especially concerned with the various way in which the many social contexts of language performance can influence interpretation. Pragmatics goes beyond such influences as supra-segmental phonemes, dialects, and registers (all of which also shape interpretation) and looks at speech performance as primarily a social act ruled by various social conventions.

Anyone who is not a hermit lives in daily contact with other human beings, learns the explicit and implicit codes by which human beings usually manage to keep from doing violence to one another, and responds to alterations in those codes with greater or lesser good nature and skill. We drive on the right-hand side of the road and expect other drivers to do so. When we write checks, we have the money to cover them, and we expect the bank to honor them. We assume that food will be forthcoming in a restaurant, haircuts in a barber shop, gasoline—maybe—from a service station. We know these contexts so well that we do not think much about them, nor do we often stop to list the expectations we have about the behavior of people in such contexts. Ordinarily, there is no need to be explicit, because ordinarily everybody else is behaving as we would expect them to. The "unspoken rules" governing behavior work very well, most of the time.

Occasionally, however, they do not work, or someone is not aware of them, or they are deliberately violated. Then it becomes important to be explicit about the "rules," the silent expectations and conventions, in order to discover what they ask of us and whether they are worth saying or not. For example, many people in the past decade or so have come to question certain "rules" about what gentlemen should do for ladies (open doors, light cigarettes, carry packages, etc.), asking whether those behaviors are fixed for all time by generic requirements or perhaps are signs, in the semiotic sense, of role-playing and strategies for coping with conflict. Similarly, pragmatics attempts to identify the "rules" underlying the performance of speech acts, or language as it is uttered in conjunction with the many social conventions controlling what speaker and auditor expect from one another.

Pragmatics theorists have identified three kinds of speech-act principles: illocutionary force, referring to the speaker as interpreted pragmatically by his auditors, conversational principles, referring to the auditors' expectations of the speaker, and presuppositions, referring to assumptions held by both the speaker and the auditors. Each of the three principles, of course, influences the others and therefore influences the significance of the speech act as a whole.

1. Illocutionary Force

Illocutionary Force
(意在言外)

This is the speaker's intention, so far as the auditors can discern it form the context. There are two major kinds of illocutionary force: implicit, below the surface and unstated, and explicit, on the surface and stated. The implicit forces are three: assertion, imperative, and question (sometimes called interrogative). Assertion is a statement about action or attitude ("He loves you," "He does not love you"). An imperative is a command for action ("Shut up!" "Will you please shut up!"). An interrogative is a request for information ("How much is that tie?" "What time is it?"). It is important to identify these implicit forces not only theoretically, but also as they appear in their various social contexts, for frequently the apparent intention of the speaker is not the same as the actual intent.

Social convention and good manners usually dictate, for instance, that a speaker will not use imperatives in polite company, perhaps at a party, at dinner, or when he is courting someone's favor. We are taught very early to say "please" as a way of disguising the illocutionary force of a command: "Please pass the biscuits"; "Give me the salt, please." It is even more polite to phrase the imperative as a question: "May I [or Can I] get through here?" "Would you like to go home now?" Most of us recognize that the implicit illocutionary force of these apparent questions is imperative, not interrogative, and we send the salt down or open a passageway without demur. We do not ordinarily respond to such implied commands by saying, "No" or by saying, "Yes" and not handing along the biscuits. We all understand that "May I have the biscuits?" is not a request for information, to be answered by "Yes" or "No," which would then have to be followed by another request for information— "Will you send them down here?"—which could then also be answered "Yes" or "No," and on and on while the biscuits stayed where they were and got cold.

Sometimes, however, the implicit illocutionary force of an utterance is not so clear, for it is often disguised by the surface-structure phrasing. When someone sitting outdoors on a cool evening says, "I'm cold," that is phrased as a statement; it apparently requires neither information nor action. But if there is a wrap in the house, brought along for just this chilliness, and if the speaker's companion is attentive, the simple statement will probably be recognized as an implied command to bring the wrap out. Similarly, the statement "You're driving too fast" (assertion) may often carry the implicit illocutionary force of a command to slow down. "Do you love me?" —an apparent question—may carry many different implicit illocutionary forces: really a question, to be answered "Yes" or "No"; an assertion (perhaps "I love you" or perhaps "I am uncertain about your love for me"); or a command ("Tell me you love me"). Only context, linguistic or otherwise, will clarify this complex utterance.

It may be tentatively suggested that the more intimate the register, the more disguised the implicit illocutionary force in any given speech act. Conversely, the more formal is the register, the less disguised the force. Drill instructors in the armed forces do not suggest; they command. Their audience is presumed to be unfamiliar with the nuance of social convention that would instantaneously translate "Why don't we go for a walk?" into "Fifty-mile forced march, full packs, on the double!" Formalized situation tend to call for formalized utterances, so that an audience of varied background does not have to fumble with unfamiliar codes and levels of implied illocutions.

The other major kind of illocutionary force is explicit. Explicit illocutionary forces in speech acts take the form of statements in which the utterance itself is an action. "I tell you, it was awful!" performs the act of telling which the verb names. "I pronounce you man and wife" performs the act of pronouncing. "I promise I'll break your head" constitutes the act of promising. Statements like these, promising or pronouncing or telling (or asking or commanding), are called performative utterances; the utterance itself is the deed. There is an understood contract in such utterances, for assertions like these always carry the force of an unspoken command. The unspoken (implicit) is that the auditor should believe the assertions to be true (should accept their truth value): it is true that something was awful; you are man and wife; your head will get broken.

Most of the time, we do accept such assertions as true, or we pretend to do so, but if the context is intimate enough, the implicit truth value may be questioned even here: "Oh, yeah? Who says it was awful? You wouldn't know 'awful' if it bit you!" Or "Oh, yeah? You and what army gonna break my head?"

(But rarely "Oh, yeah? Who says you're married?" for the context here is formal, not intimate.) When explicit and implicit intentions clash over a performative utterance, the auditors are challenging the speaker's capacity, not to tell the truth, but to verify the truth of the statements. The speaker is challenged to match the truth value of the utterance to some external referent or some action.

2. Conversational Principles

This brings us to what the auditor can expect from a speaker, as opposed to the interpretive skills that a speaker can expect from his audience. In any speech act, the audience generally assumes that at least four conversational principles will apply to what a speaker says. The audience's first assumption is that the speaker is sincere, not saying one thing and meaning another, at least with no greater discrepancy between phrasing and intention than what we expect in the exercise of various illocutionary forces. The second assumption is that the speaker is telling the truth so far as he understands it, not deliberately telling lies. Third, the audience assumes that what the speaker has to say is relevant to the topic or general areas of concern. The final assumption is that the speaker will contribute the appropriate amount of information or commentary, not withhold anything important and not rattle on for an undue amount of time.

For example, if someone (speaker) asks, "What time is it?" we (auditors) usually assume that he does not know what the time is and that his request is a sincere one for information about the time. When we (now speakers) begin to reply, he (now auditor) will usually assume that we will answer with the correct time, not with a rambling discourse on the price of hamburger, nor with a scream of rage, nor with a lie. If any of those four assumptions prove incorrect, then discord immediately appears and one or the other of the conversants has to make a quick test of the assumptions, to discover which one has been violated and what the appropriate response should now be.

For example, should people be hurrying out of a burning building, a request for the correct time is presumed to be insincere and will elicit irritation or disgust: "You crazy? Keep moving!" But if someone begins a prepared speech on tax reform to the Lions Club with a joke about peanut butter, the audience will unconsciously recognize that the relevance of the joke is less to taxes than to the reduction of stress between strangers. It is understood here that the context of speech giving requires some preliminary establishment of shared concerns, even a sort of shared companionship, between the orator and the audience. Such a speaker is not expected to launch immediately into the technical points of his topic. But if that speaker's boss asks for a short telephone conversation on the same topic of tax reform, joke would not be appropriate as an opening; it would be irrelevant in the context of a business discussion.

3.Presuppositions

Here, we move into what both speakers and auditors can expect of the content or information contained in an utterance, that is, what a speaker and an auditor can suppose each other to know before a given speech act begins. For example, if I say, "But Jenny has never gone out with a married man before!" I presuppose (before I utter the sentence) that my listener knows at least these content items:

1. There is a person named Jenny.

2. Speaker and listener are both acquainted with Jenny.

3. Jenny goes out with men.

4. Jenny has just recently gone with a married man.

Presumably, too, the listener and speaker both share the following bits of information, although these presuppositions are not so obvious nor demonstrable from the utterance alone:

5. Jenny is female.

6. Jenny is not married.

7. Jenny does not usually go out with married men.

8. Jenny is adult.

9. Jenny is not so dependent upon speaker or listener that her behavior can be regulated by either of them.

10. Speaker and listener are surprised by Jenny's behavior.

Presupposition underlies a good deal of the unthinking adjustment we make from one speech situation to another, adjustment that helps ensure we are not (as speech-communication teachers say) "talking over our audience's head" or not "insulting our audience's intelligence." Presupposition is operating when we mutter secrets in hallways so- that outsiders will not understand. All codes, jargons, cants, and deliberate use of elliptical or confusing language make use of presupposition.

It is very easy to misjudge presupposition when one does not know one's audience well, or when one thinks one knows them all too well. A good many people seem to sense this principle, as evidenced by the frequency with which they intersperse phrases like "You know," "I mean," or "Know what I mean?" in their conversation. For instance, if I tell you that I will meet you on the corner of Third and Main at noon today, I assume that you know what noon is, where Third Street makes a corner with Main Street, and which of the four points of that intersection I will be waiting on. If you are new in town, I may wait a long time. Conversely, if my husband tells me hehis favorite meal for dinner and presupposes I will cook it, because we have been married a long time and I always know and cook what he asks for, then he may wait a long time. Presuppositions always require testing from time to time, to be sure that what the speaker and what the auditor assume or know are really the same.

An attentiveness to the unspoken and often unconscious "rules" or expectations in speech acts can help to sharpen our awareness of what is really going on as we speak. The illocutionary force implicit in certain contexts, the active nature of performative statements, the conversational principles applicable to most speech situations, and the presuppositions all of us bring to conversations: these pragmatic contexts of language use shape our performance all the time. The more we understand them, perhaps the better we can control them. The same may be said of our control over individual meanings as well.

Words and Expressions

pragmatics [præg'mætiks]	n.	语用论；语用学
illocutionary [ˌiləˈkjuːʃənəri]	adj.	语言外表现行为的，以言行事，言外行为
presupposition [ˌpriːsʌpəˈziʃn]	n.	假设的事情，假定，预设
context ['kɒntekst]	n.	语境，上下文，背景
supra-segmental	n.	超音段；所有超音段
phoneme ['fəʊniːm]	n.	音位，音素

explicit	[ik'splisit]	adj.	明确的,清楚的
implicit	[im'plisit]	adj.	不言明[含蓄]的
generic	[dʒə'nerik]	adj.	类的,属性的;一般的
semiotic	[ˌsemi'ɒtik]	adj.	符号学的
discern	[di'sə:n]	vt.	看出;理解,了解;识别
assertion	[ə'sɜ:ʃn]	n.	声称;主张;论断,断定
imperative	[im'perətiv]	n.	祈使(句)的,命令,祈使语气
intent	[in'tent]	n.	意图,目的;意思,含义
dictate	[dik'teit]	vt.	指示;使听写;控制,支配
demur	[di'mə:]	n.	反对,异议
disguise	[dis'gaiz]	vt.	隐瞒,掩饰;伪装,假装
attentive	[ə'tentiv]	adj.	周到的,殷勤的;细心的
wrap	[ræp]	n.	膝毯,披肩,围巾,外套,大衣
instantaneously	[ˌinstən'teiniəsli]	adv.	突如其来地,即刻地,瞬时地
rambling	['ræmbliŋ]	adj.	闲逛(聊)的;(思想)散漫的;杂乱的
preliminary	[pri'liminəri]	adj.	初步的,初级的;预备的
presupposition	[ˌpri:sʌpə'ziʃn]	n.	假设的事情,假定,预设
jargon	['dʒɑ:gən]	n.	行话;行业术语
cant	[kænt]	n.	(下层社会的)黑话
elliptical	[i'liptikl]	adj.	椭圆的;省略的
intersperse	[ˌintə'spə:s]	vt.	散布,散置;点缀
intersection	[ˌintə'sekʃn]	n.	交叉点
applicable	[ə'plikəbl]	adj.	适当的,可应用的

Notes

1. Since each of the three speech-act principles influences the other two, it will be recognized that to separate the presupposition from the other two is to be somewhat arbitrary. The presupposition about content will vary from one speech situation to another, depending on the influence of intention and expectation.

Questions for Discussion and Review

1. One of the three basic kinds of speech act principles is illocutionary force,

a term that refers to the speaker's communicative intention to the extent that the hearers can discern. It is divided into implicit illocutionary force and explicit illocutionary force. Define these three concepts in your own words, and supply some examples of your own if necessary.

2. Heatherington suggests that "the more intimate the register, the more disguised the implicit illocutionary force in any speech act. Conversely, the more formal is the register, the less disguised the force." Do you agree? Support your answer with specific examples.

3. Do you know some fixed (conventionalized) expressions for these types of occasions(request, challenge, answer to obvious questions, reply to a false assertion)? How would you explain (to someone learning English as a foreign language, for example) the way to work out the communicated meaning from the literal meaning?

4. According to the conversational principles described by Heatherington in the text, describe a situation in which all the principles are observed and one in which one or more are violated. What conclusions can you draw?

5. Presuppositions involve both speakers and hearers; and all utterances, even the simplest, involve a number of presuppositions. Examine the following sentences and list the presuppositions for each:

a) Mary's husband works for the IBM.

b) Even though Bob promised never to lie to me again, he told me today that he didn't go to the movies with Sherry.

c) That C I got on the Psych quiz you missed is really going to hurt my average.

d) Non-smokers have rights too!

Unit 14

Psycholinguistics

> Psycholinguistics is the study of the mental processes that a person uses in producing and understanding language and how humans learn language. Psycholinguistics includes the study of speech perception, the role of memory, concepts and other processes in language use, and how social and psychological factors affect the use of language. The excerpt below is written by Herbert H. Clark, Professor of Psychology at Stanford University. It explains how speaking and listening are proceeded. It is said that speaking and listening have several levels. Speakers are supposed to plan what they say from the top level of language down—from intention to articulation while listeners are often considered to work from the bottom up—from the sounds they hear to what the speakers mean. People have to coordinate speaking with listening at many levels to accomplish what they intend to do.

Psycholinguistics is the study of people's actions and mental processes as they use language. At its core are speaking and listening, which have been studied in domains as different as LANGUAGE ACQUISITION and language disorders. Yet the primary domain of psycholinguistics is everyday language use.

Speaking and listening have several levels. At the bottom are the perceptible sounds and gestures of language: how speakers produce them, and how listeners hear, see, and identify them. One level up are the words, gestural signals, and syntactic arrangement of what is uttered: how speakers formulate utterances, and how

listeners identify them. At the next level up are communicative acts: what speakers do with their utterances, and how listeners understand what they mean. At the highest level is DISCOURSE, the joint activities people engage in as they use language. At each level, speakers and listeners have to coordinate their actions.

Speakers plan what they say more than one word at a time. In conversation and spontaneous narratives, they tend to plan in intonation units, generally a single major clause or phrase delivered under a unifying intonation contour (Chafe 1980). Intonation units take time to plan, so they often begin with pauses and disfluencies (*uh* or *um*, elongated words, repeated words). For example, one speaker recounting a film said: "[1.0 sec pause] A-nd u-m [2.6 sec pause] you see him taking…picking the pears off the leaves."

Planning such units generally proceeds from the top level of language down—from intention to ARTICULATION (Levelt 1989). Speakers decide on a message, then choose construction for expressing it, and finally program the phonetic segments for articulating it. They do this in overlapping stages.

Formulation starts at a functional level. Consider a woman planning "Take the steaks out of the freezer." First she chooses the subject, verb, direct object, and source she wants to express, roughly "the addressee is to get meat from a freezer." Then she chooses an appropriate syntactic frame, an imperative construction with a verb, object, and source location. She then finds the noun and verbs she needs, *take*, *steak*, and freeze. Finally, she fills in the necessary syntactic elements—the article *the*, the preposition *out of*, and the suffixes -*s*, and -*er*. Formulation then proceeds to a positional level. She creates a phonetic plan for what she has formulated so far. She uses the plan to program her articulatory organs (tongue, lips, glottis) to produce the actual sounds, "Take the steaks out of the freezer." Processing at these levels overlaps as she plans later phrases while articulating earlier ones.

Much of the evidence for these stages comes from slips of the tongue collected over the past century (Fromkin 1973; Garrett 1980). Suppose that the speaker of the last example had, by mistake, transposed *steak* and *freeze* as she introduced them. She would then have added -*s*: to *freeze* and -*er* to *steak* and produced "Take the freezes out of the steaker." Other ships occur at the positional level, as when the initial sounds in *left hemisphere* are switched to form *heft lemisphere*.

Listeners are often thought to work from the bottom up. They are assumed to start with the sounds they hear, infer the words and syntax of an utterance, and, finally, infer what the speakers meant. The actual picture is more complicated. In everyday conversation, listeners have a good idea of what speakers are trying to do, and working top down, they use this information to help them identify and understand what they hear (Tanenhaus and Trueswell 1995).

Spoken utterances are identified from left to right by an incremental process of elimination (Marslen-Wilson 1987). As listeners take in the sounds of "elephant," for example, they narrow down the words it might be. They start with an *initial cohort* of all words beginning with "e" (roughly 1000 words), narrow that to the cohort of all words beginning with "el" (roughly 100 words), and so on. By the sound "f" the cohort contains only one word, allowing them to identify the word as "elephant." This way listeners often identify a word before it is complete. Evidence also suggests that listeners access all of the meanings of the words in these cohorts (Swinney 1979). For example, the moment they identify "bugs" in "He found several bugs in the corner of his room" they activate the two meanings "insects" and "hidden microphones." Remarkably, they activate the same two meanings in "He found several spiders, roaches, and other bugs in the corner of his room," even though the context rules out microphones. But after only 2 to 4 seconds "hidden microphones" gets suppressed in favor of "insects."

Still, listeners do use top-down information in identifying words and constructions (Tanenhaus et al. 1995). When people are placed at a table with many objects on it and are asked, "Pick up the candle," they move their gaze to the candle before they reach for it. Indeed, they start to move their eyes toward the candle about 50 msec before the end of "candle." But if there is candy on the table along with the candle, they do not start to move their eyes until 30 msec after the end of "candle." As a sentence, "Put the apple on the towel in the box" may mean either (1) an apple is to go on a towel that is in a box, or (2) an apple on a towel is to go into a box. Without context, listeners strongly prefer interpretation 1. But when people are placed at a table with two apples, one on a towel and another on a napkin, their eye movements show that they infer interpretation 2 from the beginning. In identifying utterances, then, listeners are flexible in the information they exploit—auditory information, knowledge of syntax, and the context.

Speaking and listening aren't autonomous processes. People talk in order to do things together, and to accomplish that they have to coordinate speaking with listening at many levels (Clark 1996).

One way people coordinate in conversation is with adjacency pairs. An adjacency pair consists of two turns, the first of which projects the second, as in questions and answers:

Sam: And what are you then?
Duncan: I'm on the academic council.

In his first turn Sam proposes a simple joint project, that he and Duncan exchange information about what Duncan is. In the next turn Duncan takes up his proposal, completing the joint project, by giving the information Sam wanted. People use adjacency pairs for establishing joint commitments throughout conversations. They use them for openings (as in the exchange "Hey, Brabara" "Yes?") and closings ("Bye" "Bye"). They use them for setting up narratives ("Tell you who I net yesterday—" "Who?"), elaborate questions ("Oh there's one thing I wanted to ask you" "Mhm"), and other extended joint projects.

Speakers use their utterances to perform illocutionary acts—assertions, questions, requests, offers, promises, apologies, and the like—acts that differ in the uptake they project. Most constructions (e. g., "Sit down") can be used for more than one illocutionary act (e. g., a command, a request, an advisory), so speakers and listeners have to coordinate on what is intended. One way they coordinate is by treating each utterance as a contribution to a larger joint project. For example, when restaurant managers were asked on the telephone, "Do you accept American Express card?" they inferred that the caller had an American Express card and wanted a "yes" or "no" answer. But when they were asked "Do you accept any kinds of credit card?" they inferred the caller had more than one credit card and wanted a list of the cards they accepted ("Visa and Mastercard"). Listeners draw such inferences more quickly when the construction is conventionally used for the intended action. "Can you tell me the time?" is a conventional way to ask for the time, making it harder to construe as a question about ability (Gibbs 1994).

People work hard in conversation to establish that each utterance has been understood as intended (Clark 1996). To do that, speakers monitor their speech for problems and repair them as quickly as reasonable (Levelt 1983; Schegloff, Jefferson, and Sacks 3977). In "if she'd been—he'd been alive," the speaker discovers that "she" is wrong, replaces it with "he," and continues. Listeners also monitor and, on finding problems, often ask for repairs, as Barbara does here:

Alan: Now-um do you and your husband have a j-car?
Barbara: Have a car?
Alan: Yeah.
Barbara: No.

People monitor at all levels of speaking and listening. Speakers, for example, monitor their addressees for lapses of attention, mishearings, and misunderstandings. They also monitor for positive evidence of attention, hearing, and understanding, evidence that addressees provide. Addressees, for example, systematically signal their attention with eye gaze and acknowledge hearing and understanding with "yeah" and "uh huh."

Speaking and listening are not the same in all circumstances. They vary with the language (English, Japanese, etc.), with the medium (print, telephones, video, etc.), with age (infants, adults, etc.), with the genre (fiction, parody, etc.), with the trope (irony, metaphor, etc.), and with the joint activity (gossip, court trials, etc.). Accounting for these varitions remains a major challenge for psycholinguistics.

Words and Expressions

perception [pəˈsepʃn]	n.	知觉；感知
domain [dəˈmein]	n.	范围,领域
perceptible [pəˈseptəbl]	adj.	可感觉[感受]到的
spontaneous [spɒnˈteiniəs]	adj.	自发的；自然的；无意识的
contour [ˈkɒntʊə(r)]	n.	外形,轮廓
overlap [ˌəʊvəˈlæp]	n.	重叠部分 vt/ vi. 互搭,重叠
transpose [trænˈspəʊz]	vt /vi.	变换顺序,进行变换
incremental [ˌiŋkrəˈmentl]	adj.	增量的；渐进,增加的
cohort [ˈkəʊhɔːt]	n.	步兵大队,军队；队列
roach [rəʊtʃ]	n.	蟑螂

autonomous	[ɔːˈtɒnəməs]	adj.	自治的；有自主权的；自发的
adjacency	[əˈdʒeisnsi]	n.	邻接；邻近；邻接关系
lapse	[læps]	n.	小错，疏忽；行为失检，失足
addressee	[ˌædreˈsiː]	n.	收信人，收件人
with the genre	[ˈʒɑnrə]	n.	类型，种类；体裁，样式
parody	[ˈpærədi]	n.	拙劣的模仿；恶搞

Questions for Discussion and Review

1. Speaking and listening are considered to have several levels. Explain the processes of speaking and listening at different levels.

2. How do people coordinate speaking with listening at many levels to accomplish what they intend to do?

Unit

Artificial intelligence and computer modeling

It is sometimes argued (e. g. Sloman, 1978, 18) that psychological experiments are superfluous at the present time, because so many everyday observations remain unexplained. In the field of language understanding, for example, it is apparent that people understand text, yet there is no adequate theory of how they do so. More specifically no experiments are needed to show that people can readily determine when two expressions (for example, the man and he) refer to the same person or object, or that they can recognize when an action of a character in a story is an attempt to achieve a certain goal, yet it is not possible at present to say how such judgements are made. Given the assumption that these abilities are amenable to scientific study, it should be possible to model them in detail. However, psychologists, at least according to their detractors, have never attempted to provide such an account of how understanding is achieved. Many practitioners of AI (e. g. Sloman, 1978) have argued that before any experiments are performed to test between psychological theories of language processing, it is necessary to formulate a sufficiently detailed theory. They point out that if a computer were programmed to simulate language understanding, then, because of the nature of computers, every step in the comprehension process would have to be explicitly specified. Therefore a computer program for understanding language would automatically embody a theory of how comprehension could be achieved. It may not comprehend in the way that people do, but it would demonstrate the form that an adequate theory might take. It is sometimes further argued that because comprehension is such a complex process, it can be achieved in only a limited number of ways, which must all be similar to one another. Hence writing a computer program to understand language would be a major step towards a theory of how people comprehend.

165

Another argument for writing programs is that a program that simulated language understanding would have to include models of all the subcomponents of the processing system, and incorporate some idea of how they act together. The problem of what happens when the bits are combined cannot be ignored.

The main advantage of computer programs over most psychological theories is that a program must be, in some sense, formally precise. High-level programming language, many of them specially developed for AI, provide building blocks for well-specified theories. Indeed many psycholinguists have recognized the usefulness of AI formalisms and incorporated them into their own theories. Linguistic formalisms are useful for a similar reason. Unfortunately the formal apparatus of programming languages does not, of itself, guarantee an insightful, or even a testable, theory of language understanding. Furthermore, some AI research goals have worked against the production of psychologically interesting theories.

Some of these goals stem from the fact that AI is in many ways more similar to engineering than pure science. The aim of an AI research project is often to produce an 'intelligent' machine with a practical application. Such a project is successful if it results in the writing of a program that produces realistic outputs. Even when no applications are likely, this measure of success is often used in AI. However, a working program is of psychological interest only if it is based on general explanatory principles about the way the mind works. It is difficult to deduce such principles from programs themselves, whose creators have often omitted to formulate them. The problem is made worse by the fact that, in order to make programs work it is usually necessary to include sections, called *patches*, to perform parts of the task that the programmer is either not interested in, or has no theory about. Such *ad hoc* simulation does not provide any kind of explanation, and obscures the way in which underlying principles contribute to the program's output. Furthermore, as Weizenbaum's (1996) ELIZA program, which simulates a Rogerian (non-directive) psychotherapist, demonstrates, realistic outputs do not indicate that any theoretically useful analysis of language understanding has been made.

AI's potential contribution to psycholinguistics is a set of theoretically interesting principles that govern the operation of the language understanding system. The usefulness of programming comes not from the fact that its end result is a program that can 'talk', but from the fact that, when programmers try to produce a principled model of a linguistic subprocessor, and not a patch to make a program work, they are forced to think carefully about the sequence of operations that the subprocessor performs to produce its output, which might be, for example, a (partial) parse tree. This approach is very different from that of the experimental psychologist, who simply looks for an effect of the manipulation of variables.

1.Programs and theories

Occasionally an AI worker (e. g. Schank,1973)has claimed that a computer program is itself a scientific theory of, say, the way in which people understand language. A more plausible view (e. g. Isard,1974;Johnson-Laird,1982)is that a program is a model of a theory. Unfortunately, it is comparatively rare to find the theory set out in a standard scientific way with its principal tenets stated in a concise form. The result is that the theory embodied in a program is often difficult to grasp, and hence of limited use in explaining psychological facts. However, sometimes the theoretical claims are clear. For example, the theory might make a claim about

flying robot

the format in which linguistic rules are stored, and the program may contain many instances of rules stored in that way. Or the theory might state that a procedure of a certain kind is used to compute, say, surface syntactic structure from the rules of the grammar and the input sentence, and that procedure will be written into the program in a certain way. Pure, as opposed to applied, AI research should provide a set of principles that govern the operation of the language operating system. Working programs are useful for showing that such principles work in practice, but it is important that the principles themselves should be formulated as clearly and precisely as the program (cf. VanLehn, Brown and Greeno, 1984)

2.Empirical tests of programs

As has already been mentioned, most AI researchers test their programs by ensuring that they produce realistic outputs. The programs are then evaluated in terms of the range of inputs they can process, and how satisfactory their responses are. However, it is not necessary to model the whole of the language understanding system in order to test psycholinguistic theories. The main argument put forward for producing such global models was that, since the various parts of the understanding system interact with one another in comprehension, it might be misleading to study any one component in isolation. The evidence for interaction is not now as strong as it was when this argument was originally put forward in the early 1970s, but even if it were, the argument would still be fallacious. If the operation of one processor is affected by the output of another, only the output of that second processor need be known in order to test a model of the first. It is not necessary to know how the second processor works. Thus, even if the language processor is interactive, a more standard scientific approach can be adopted. The problem of understanding language processing can be split up into subproblems, and models of individual subprocessors can be tested, provided that two conditions are met. First, there must be a specification of the input that the processor receives, and the output that it produces. For example, the word recognition system receives as input perceptual information, and perhaps the output of other subprocessors. Its output is the identity of the word currently being examined. Second, relevant outputs of other processors must be known. These outputs maybe identified either from common sense, or by appeal to linguistic theory. The problem of identifying what subprocessors the language understanding system contains is a difficult, but separate, issue.

3. Cognitive science—a synthesis

Neither experimental psychology nor AI provides a wholly satisfactory approach to the study of language understanding. In recent years many members of the two research communities have become aware of the advantages that the other has to offer. A new synthesis of the two approaches—cognitive science—is beginning to emerge, combining the best points of the two. Cognitive science includes not only cognitive psychology and AI, but also linguistics and philosophy, from which psychology and AI have both formalisms and rules. Cognitive science recognizes the importance of formalisms and the importance of giving a detailed account of language understanding, but it eschews the idea of programming for programming's sake. It draws clear distinctions between pure and applied science, and between formulating underlying principles and solving practical problems using existing resources. At the present time the way forward in psycholinguistics appears to be the way of cognitive science.

4. Summary

Psycholinguistics aims to understand the mechanisms of language use. For practical reasons there has been a strong bias towards studying comprehension rather than production, and, in some areas, a bias towards written rather than spoken language.

Language is about the world—or sometimes about a fictitious world—and the primary tasks of the understanding system are to work out first the situation that a particular linguistic input is about, and second, the point of that input—description, question, command, promise or whatever. Thus the understanding system must construct a representation of a part of the world corresponding to the current discourse or text—a mental model, and work out what it must do with that model. Several subprocesses contribute to this operation—word recognition, parsing, semantic interpretation, model construction and pragmatic interpretation. A theory of language understanding should provide models of the subprocessors responsible for carrying out these tasks, and an account of the way in which they act together to effect understanding.

There are two main approaches to the study of language understanding—that of experimental psychology and that of AI. Both approaches have their strengths and defects, and a combination of the strengths of the two is found in cognitive science. AI provides the formalism for constructing well-specified theories, and the idea that every detail of the process of language understanding should ultimately be spelled out.

Words and Expressions

superfluous [suːˈpəːfluəs]	adj.	过多的；多余的；不必要的
be amenable to	v.	有义务，顺从，经得起检验
detractor [diˈtræktər]	n.	贬低者；诋毁者

practitioner [præk'tiʃənɚ]	n.	从业者;实践者
explicitly [ik'splisitli]	adv.	明白地,明确地
embody [im'bɒdi]	vt.	表现,象征;包含
simulate ['simjuleit]	vt.	模仿;模拟;仿生
incorporate [in'kɔːpəreit]	vt/vi.	包含;吸收;合并;混合
formalism ['fɔːməlizəm]	n.	形式主义
application [ˌæpli'keiʃn]	n.	适用,应用,运用;申请表
deduce [di'djuːs]	vt.	推论,推断;演绎
patche [pætʃ]	n.	补丁,眼罩;斑点;小块
ad hoc [ˌæd 'hɒk]	adj.	特别的;临时的;特设的
psychotherapist [ˌsaikəʊ'θerəpist]	n.	精神(心理)治疗医师
subprocessor [sʌbp'rəʊsesər]		辅助处理机,子处理
parse tree [pɑs triː]		(语法)分析树
plausible ['plɔːzəbl]	adj.	貌似真实(有理)的;花言巧语的;有眉有眼
tenet ['tenit]	n.	原则;信条;宗旨
format ['fɔːmæt]	n.	(数据安排的)形式;版式,格式化;格式
Empirical [im'pirikl]	adj.	经验主义的;以观察或实验为依据的
fallacious [fə'leiʃəs]	adj.	谬误的;欺骗的;靠不住的
processor ['prɑːsesə]	n.	(计算机)中央处理器,加工程序;处理机
synthesis ['sinθəsis]	n.	综合体
bias ['baiəs]	n.	偏见
fictitious [fik'tiʃəs]	adj.	虚构的,假想的
parsing, ['pɑːziŋ]	n.	分[剖]析,句法分析

Questions for Discussion and Review

1. Define psycholinguistics and explain its overall goal and subgoals.

2. What does the author think are some of the biases in psycholinguistic research?

3. Discuss where psycholinguistic theories come from?

4. What methods are often used in psycholinguistic research?

5. What is cognitive science? Why is it necessary to combine experimental psychology and AI?

参 考 文 献

1. Edward Sapir. Language: An introduction to the Study of Speech. New York: Harcourt. Brace and World, Inc. 1949.
2. W. F. Bolton. Language Introductory Readings. New York: St. martin's Press. 1981.
3. Dwight Bolinger. Aspects of Language. Harcourt Brace Jovanovich, Inc. 1975.
4. Winfred P. Lehmann. Descriptive Lingristics: An introduction. Random House. Inc. 1976.
5. Ferdinand de Saussure. Course in General Lingusistics. New York: McGraw-Hill Book Company. 1959.
6. G. N. The MIT Encyclopedia of the Cognitive Sciences. Cambridge. Mass.: MIT Press. 1999.
7. Robert E. Callary. Language Introductory Readings. New York: St. martin's Press. 1981.
8. Noam Chomsky. Language and Mind. New York: harcourt Brace Jovanovich, Inc. 1972.
9. Ronnie Cann(1993). Formal Semantics: An Introduction. Cambridge: Cambridge University Press. 1993.
10. Madelon E. Heatherington. Language introductory readings. New York: St. Martin's Press. 1981.
11. Herbert H. Clark. Psychololinguistics. The MIT Encyclopedia of the Cognitive Sciences. Cambridge, Mass.: MIT Press. 1999.
12. Alan Garnham, Psycholinguistics: Central topics. London: Methuen. 1985.